SECOND EDITION

How to Pass

D1326630

NATIONAL 5

Physics

Paul Chambers and
Hugh McGill

HODDER
GIBSON
AN HACHETTE UK COMPANY

The Publishers would like to thank the following for permission to reproduce copyright material:

Photo credits

p.108 © NASA - Yuri Arcurs - Fotolia.com; **p.112** © maxime SCHAAL - Fotolia; **p.114** © Carol Buchanan / Fotolia.com.

Acknowledgements

Every effort has been made to trace all copyright holders, but if any have been inadvertently overlooked the Publishers will be pleased to make the necessary arrangements at the first opportunity.

Although every effort has been made to ensure that website addresses are correct at time of going to press, Hodder Gibson cannot be held responsible for the content of any website mentioned in this book. It is sometimes possible to find a relocated web page by typing in the address of the home page for a website in the URL window of your browser.

Hachette UK's policy is to use papers that are natural, renewable and recyclable products and made from wood grown in well-managed forests and other controlled sources. The logging and manufacturing processes are expected to conform to the environmental regulations of the country of origin.

Orders: please contact Bookpoint Ltd, 130 Park Drive, Milton Park, Abingdon, Oxon OX14 4SE. Telephone: (44) 01235 827827. Fax: (44) 01235 400454. Lines are open 9·00–5·00, Monday to Saturday, with a 24-hour message answering service. Visit our website at www.hoddereducation.co.uk. Hodder Gibson can also be contacted at hoddergibson@hodder.co.uk

First published in 2018 by
Hodder Gibson, an imprint of Hodder Education,
An Hachette UK Company
211 St Vincent Street
Glasgow G2 5QY
Impression number   2

Year   2019

Cover photo © Makistock/stock.adobe.com
Illustrations by Aptara, Inc.
Typeset in 13/15 Cronos Pro (Light) by Aptara, Inc.
Printed in India
A catalogue record for this title is available from the British Library
ISBN: 978 1 5104 2105 9

Contents

Welcome to this revision book!

And welcome to National 5 (N5) Physics! This book is designed to help you.

N5 Physics is enjoyable, interesting and relevant to life in the twenty-first century. It is also very useful if you want to progress to Higher Physics, apply to college or get a job.

The examination is fair. Prepare properly for it and you will do well.

Good luck!

How to use this book

This book covers everything you need for N5 Physics. It also includes tips and hints about how to answer exam questions and about what SQA expects you to do.

The Introduction covers skills that you need for the course – general things like units, prefixes and scientific notation. It also includes advice on how to prepare for the exam. You should refer to this chapter often while you are preparing for your prelim and for the final exam.

The next six chapters cover the knowledge and understanding of the six sections detailed in the Course Specification.

Each chapter is written in topics. At the start of each topic there is a summary of what you need to be able to do. You can use these summaries to help you prepare a study checklist for your prelim or for the final examination.

Study the topics one at a time – in any order you like. Studying for long periods without any breaks is not an effective way of learning. It is better to study intensively for short periods, taking regular breaks to give your brain time to relax.

Some topics include examples that show you how to answer exam questions. All of the topics in Chapters 1 to 6 include questions for you to try. Just like in the final examination, some of the questions are straightforward and some are more difficult.

Answers to all of the questions are provided in the Solutions to exercises section at the back of the book. For maximum benefit, try to answer the questions without looking at the answers. If you do not know how to tackle a question, the answers are there to help you. Use the answers wisely and you will learn a lot.

Study all the topics and try all the questions before you sit the final examination.

Units, prefixes and scientific notation

For N5 Physics you need to be able to:
- use the SI units of all the physical quantities included in the course
- give answers to calculations to an appropriate number of figures
- understand and use scientific notation
- understand and use the prefixes n, μ, m, k, M and G.

Units

You have to be able to use the International System of Units (SI) of all the physical quantities included in the N5 Physics course. A full list of these quantities is included in the table at the end of this section.

In N5 Physics, there is a mark for the **final answer** to a numerical problem. To gain this mark your final answer must have the correct number and the correct unit for the quantity. Write your numerical answers in the form:

quantity = number and correct unit (for example, mass = 2·3 kg)
OR symbol = number and correct unit (for example, m = 2·3 kg)

It may not seem much, but making sure you get the marks for final answers can soon make the difference between passing and failing.

You do **not** have to write units in the middle of calculations, but your statement of the final answer **must** include the correct unit.

Significant figures

The number of figures in your **final answer** to a numerical question (or part of a question) should be the same as the **minimum** number of significant figures given in the data that you use to work out the answer.

Do not round intermediate values to the number of significant figures in the question. Keep one additional figure in intermediate values and **round only when you write your final answer**. This method is used in the examples and in the solutions to the exercises in this book.

Your examination booklet will contain a Data Sheet. Be careful when you use data from the Data Sheet as this data may have fewer significant figures than data given in the question.

Having too many or too few figures in final answers can cost you marks – do not let this be the reason you do not pass your N5 Physics!

Scientific notation

Some of the data used in this course are very large or very small numbers. You have to understand the notation for these numbers and be able to use it correctly.

For example, speed of light in a vacuum = $3·0 \times 10^8$ m s^{-1}.

Make sure you practise using scientific notation often!

Prefixes

The table below includes all the prefixes that you need to understand for N5 Physics. Learn them and get into the habit of using them.

Prefixes used in N5 Physics

Prefix	Short for	Means	Prefix	Short for	Means
m	milli	$\times 10^{-3}$	k	kilo	$\times 10^{3}$
μ	mu	$\times 10^{-6}$	M	mega	$\times 10^{6}$
n	nano	$\times 10^{-9}$	G	giga	$\times 10^{9}$

The kilogram is the SI unit for mass – **do not** change kilograms to grams.

Physical quantities in National 5 Physics

The table below shows all the quantities that you will meet in N5 Physics, with their symbols, units and abbreviations.

Physical quantities included in the N5 Physics course

Physical quantity	Symbol	Unit	Abbreviation
area	A	metre squared	m^2
volume	V	metre cubed	m^3
distance, displacement	s or d	metre	m
height	h	metre	m
time	t	second	s
speed, instantaneous speed, velocity, final velocity	v	metre per second	$m\,s^{-1}$
initial velocity	u	metre per second	$m\,s^{-1}$
change of velocity	Δv	metre per second	$m\,s^{-1}$
average velocity, average speed	$\overline{v}$	metre per second	$m\,s^{-1}$
acceleration	a	metre per second per second	$m\,s^{-2}$
acceleration due to gravity	g	metre per second per second	$m\,s^{-2}$
gravitational field strength	g	newton per kilogram	$N\,kg^{-1}$
mass	m	kilogram	kg
force, resultant force	F	newton	N
weight	W	newton	N
energy	E	joule	J
work done	E_w	joule	J
potential energy	E_p	joule	J
kinetic energy	E_k	joule	J
power	P	watt	W
percentage efficiency	-	-	-
output energy	E_o	joule	J
input energy	E_i	joule	J
output power	P_o	watt	W
input power	P_i	watt	W

(cont.) Physical quantities included in the N5 Physics course

Physical quantity	Symbol	Unit	Abbreviation
heat energy	E_h	joule	J
temperature	T	degree Celsius	°C
specific heat capacity	c	joule per kilogram per degree Celsius	$J\,kg^{-1}\,°C^{-1}$
specific latent heat	l	joule per kilogram	$J\,kg^{-1}$
pressure	P	pascal	Pa
electric charge	Q	coulomb	C
electric current	I	ampere	A
voltage, potential difference	V	volt	V
resistance	R	ohm	Ω
total resistance	R_T	ohm	Ω
supply voltage	V_S	volt	V
resistance of resistor	$R_1, R_2...$	ohm	Ω
voltage across resistor	$V_1, V_2...$	volt	V
number of turns in primary coil	n_p	-	-
number of turns in secondary coil	n_s	-	-
voltage across primary coil	V_p	volt	V
voltage across secondary coil	V_s	volt	V
current in primary circuit	I_p	ampere	A
current in secondary circuit	I_s	ampere	A
voltage gain	V_{gain}	-	-
output voltage	V_o	volt	V
input voltage	V_i	volt	V
power gain	P_{gain}	-	-
period	T	second	s
frequency	f	hertz	Hz
wavelength	λ	metre	m
activity	A	becquerel	Bq
number of nuclei decaying	N	-	-
absorbed dose	D	gray	Gy
radiation weighting factor	W_R	-	-
equivalent dose	H	sievert	Sv
half-life	$t_{\frac{1}{2}}$	second	s

Preparing for the exam

On the day of your N5 Physics exam you need to:
- be aware of what the exam involves
- know how to answer multiple-choice questions
- present answers to numerical questions so that you gain as many marks as possible
- write **clear**, **understandable** and **relevant** descriptions, explanations and conclusions
- manage your time so that you complete the whole paper.

The N5 Physics exam

The N5 Physics exam lasts for **2 hours and 30 minutes** and is marked out of a total of **135 marks**. Questions will be spread across all six content areas of the course: Dynamics, Space, Electricity, Properties of matter, Waves and Radiation.

At the start of the paper there are multiple-choice questions worth a total of 25 marks. The remaining 110 marks require responses of a few words or sentences, or numerical calculations. The score for the 110-mark section is scaled to 75 and this is added to your multiple-choice score to give a total out of 100.

The majority of the marks will be awarded for applying your knowledge and understanding. The remaining marks will be awarded for applying scientific enquiry, scientific analytical thinking and problem solving.

Key points

Knowledge and understanding questions test your ability to:
- use quantities and units
- use relationships to solve straightforward numerical problems
- apply Physics principles in familiar situations
- describe familiar models, for example the nuclear model of the atom.

It is very important that you gain as many marks as possible for the knowledge and understanding questions. This will give you a solid basis for passing the exam and achieving a good grade.

You will only be able to get these marks if you prepare carefully for the exam – so **study properly!**

Key points !

Questions on problem solving, scientific enquiry and analytical thinking and skills test your ability to:

* select and present **relevant** information in experimental and other contexts
* solve numerical questions in a variety of contexts
* draw **valid** conclusions from information
* explain observations
* plan, design and evaluate experimental procedures
* integrate skills across the course.

Some of the problem solving questions may be set in reasonably familiar situations.

The remaining questions may be set in contexts that you have never seen before. These questions are generally the most difficult in the exam. **Do not panic when you find questions like this**. You can easily pass your N5 without getting marks for these questions. You will need to get marks in these questions if you want to get a grade A, however.

Multiple-choice questions

Do not expect all the multiple-choice questions to be easy. Some will be straightforward and some will be difficult.

In N5 Physics there is **one** correct answer to each multiple-choice question. There are also **four wrong answers** designed to distract you. A good way to avoid being distracted is to cover the possible answers while you read and think about the question.

If you can, work out an answer before looking at the possible answers given in the question. If you are not sure, improve your chances by eliminating the answers that you know are definitely wrong. If you are completely stuck, guess.

Do not spend too long on the multiple-choice questions – about 20–25 minutes should leave you enough time to complete the rest of the paper.

Numerical questions

When you are doing numerical questions, follow the steps below – this will help you to get as many marks as possible.

1 Collect data in a column on the left-hand side of the page. Use the symbols used by SQA and include the symbol for the quantity you want to find (remember, you may need data from the Data Sheet).
2 Check the units of each piece of data – make sure all have correct SI units.
3 Select the correct relationship from the Relationships Sheet – look for a relationship that has the symbols included in the list you made at step 1.

4 Write the relationship in the centre of the page.

5 On the line below, substitute in one number at a time. It is better to substitute numbers before rearranging the formula. Make sure the equals signs are in line one above the other (this makes it easier for you to check your working).

6 Carry out the calculation. Do this in the space at the right-hand side of the page or on your calculator.

7 Write down your answer in the form 'symbol = number'. For example, $V = 2.5$. Again, make sure the equals signs are one above the other.

8 If necessary, round to the correct number of significant figures (keep one extra figure in intermediate values).

9 Add the correct SI unit for the quantity you have calculated.

The above method is illustrated in the examples below.

Examples

How to present the answers to numerical problems

1 A cyclist moving at $3.1\,\mathrm{m\,s^{-1}}$ pedals faster for 8.0 seconds and reaches a speed of $4.3\,\mathrm{m\,s^{-1}}$. Calculate her acceleration.

$u = 3.1\,\mathrm{m\,s^{-1}}$ (**Collect** the data in a column using symbols.)

$v = 4.3\,\mathrm{m\,s^{-1}}$ (**Check** that the units are correct SI units.)

$t = 8.0$ seconds

$a = ?$ (**Include** the symbol for the quantity to be calculated.)

        (**Which relationship** includes these symbols?)

$$a = \frac{v - u}{t}$$ (Correct relationship.)

$$\Rightarrow a = \frac{4.3 - 3.1}{8.0}$$ (**Substitute** carefully.)

$$\Rightarrow a = 0.15\ \mathrm{m\,s^{-2}}$$ (Do the **arithmetic**.)

(**Check** that the final answer has the correct **unit** and number of **significant figures**.)

2 The potential difference across a $4.8\,\mathrm{k\Omega}$ resistor is $12\,\mathrm{V}$. Calculate the current in the resistor.

$R = 4.8\,\mathrm{k\Omega} = 4800\,\Omega$ (Collect data and check units are correct SI units.)

$V = 12\,\mathrm{V}$

$I = ?$ (Which relationship has R, V and I?)

$V = IR$ (Correct relationship.)

$\Rightarrow\quad 12 = I \times 4800$ (Substitute carefully.)

$\Rightarrow\quad I = 2.5 \times 10^{-3}\,\mathrm{A}$ (Do the arithmetic.)

        (Remember – significant figures and units!)

Descriptions, explanations and conclusions

In Section 2 of your N5 Physics exam paper, the number of marks for each question, or part of a question, is shown in **bold** on the right-hand side of the page. The more marks there are, the more you have to write. Normally each **relevant** piece of Physics is worth 1 mark. So, to get full marks in a 2-mark question you should include **two** relevant pieces of information.

In descriptions, explanations and conclusions the information you present must be **relevant**, **clear** and **complete**. After you have written an answer, **read it** to see if it makes sense.

In most explanation questions you need to **apply a Physics principle or relationship**. For example, questions about moving objects often involve Newton's Laws. Electricity questions often involve current and voltage rules for series and parallel circuits.

Pay attention to verbs in questions. Here are some common verbs used in SQA Physics examinations and what they mean:

- state – give a specific value or observation – no explanation required
- describe – give an outline of what happens in a given context
- explain – give a reason or reasons for an observation
- explain in terms of – use the given principle or physical quantity to explain
- explain what is meant by – define a term or phrase
- what would happen if – use a principle or relationship to make a prediction
- justify your answer – give a reason or reasons for a response you have made.

Some questions have two verbs – watch out for these. For example, *describe and explain* is not the same as *describe* on its own. Make sure your answer addresses both verbs.

Use the language of Physics. There are many terms in Physics that have precise meanings – use these words in your written responses. Paraphrasing often introduces inaccuracy and may cost you marks.

Managing your time

This is a skill that you must practise – not just for Physics but for all your subjects. Every time you sit a class test, a prelim or an exam you can practise managing your time. Check your watch (or the clock in the exam room) every 15 minutes or so. You must make sure that you do not run out of time and that you try to answer all the questions.

1.1
Electrical concepts

What you should know

For N5 Physics you need to be able to:

★ understand, define and use the terms charge, conductor, insulator, electric field, current, circuit, series, parallel and potential difference (voltage)
★ describe the path followed by positively or negatively charged particles in given electric fields
★ understand and use the relationships $Q = It$ and $E_w = QV$
★ draw and identify common circuit symbols
★ draw and interpret circuit diagrams.

Charge

Charge is a fundamental property of the particles in the Universe. For example, an electron is negatively charged and a proton is positively charged. A neutron is neither positive nor negative – it is electrically neutral. Charge is a scalar. The SI unit of charge is the **coulomb** (C) and the symbol for charge used by SQA is Q.

Key points

＊ There are two kinds of charge – positive and negative.
＊ Materials with unlike charges are attracted to each other. Materials with like charge are repelled by each other.

A balloon becomes charged when it is rubbed; when the charged balloon is held against a neutral wall it causes 'charge separation' – a slight movement of the charges in the wall molecules – like charges in the wall are repelled away from the surface of the wall and opposite charges are attracted towards the surface of the wall. As a result the attractive force between the balloon and the wall is greater than the repulsive force and the balloon sticks to the wall. The wall remains neutral throughout.

Electric fields

An electric field is a region of space where charges experience an electrical force. Learn this definition as you could be asked to state it. To draw an electric field you must show the position and charge (+ or −) of the object or objects that have created the field. You must also show the **field lines**. Field lines show the shape and direction of the electrical forces in the fields.

The field lines also show the path that a charged particle will follow in the field.

Key points !

* The direction of an electric field is the direction of the force experienced by a **positive** charge in the field.
* A **negative** charge experiences a force in the opposite direction to the direction of the field.

Figure 1.1 shows the shapes of two electric fields.

(a)

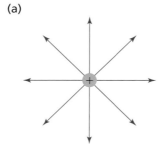

(b)

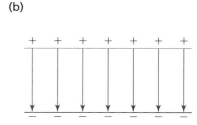

Figure 1.1 (a) Electric field around a positive point charge; **(b)** Electric field between two charged plates

Examples 🚩

1 Look carefully at the shape of the field lines around the positive point charge in Figure 1.1 (a).
   - An electron directly below the point charge is attracted towards the point charge – so the electron will move straight up.
   - A proton to the left of the point charge will move further to the left as it is repelled by the positive point charge.

2 An alpha particle in the centre of the electric field in Figure 1.1 (b) will move down as it is repelled by the positive plate and attracted by the negative plate.

Understanding the electric field between charged plates will help you to understand capacitors. You will find out more about capacitors in Topic 1.6.

Current

When an electrical force in a conductor causes charge carriers to move, we describe the flow of charge as an electric current. Current is defined as quantity of charge transferred per unit time.

One ampere is equal to one coulomb per second.

Conductors and insulators

A conductor is a material in which charge carriers can move through the material. In solids, charge carriers are usually electrons. Metals like iron, copper, brass and aluminium are good conductors. Graphite (carbon) is also a good conductor.

In liquids, charge carriers are usually positive and negative ions of dissolved materials. Water is also a conductor – this is why it is very dangerous when water comes into contact with mains electricity.

An insulator is a material in which charge carriers cannot move through the material. Glass, polythene and rubber are good insulators.

Charge flows when there is a complete path of conductors and a source of electrical energy (a supply). The complete path of conductors is called an electrical **circuit**. When a supply is switched on it applies an electric field to the charge carriers in a circuit. The electrical force causes charge carriers to flow and produces current in the conductors.

Key points

* The SI unit of current is the **ampere** (A).
* The symbol for current used by SQA is I.
* The relationship between charge, current and time is $Q = It$.

Example

An electric torch is switched on for 30 minutes. The current in the bulb of the torch is 0·45 A. Calculate the charge that flows through the bulb.

$I = 0·45\,A$ $Q = It$

$t = 30\ \text{minutes} = 1800\,s$ $= 0·45 \times 1800$

                                        $= 810\,C$

Hints & tips

Don't just use the values you're given in the question without thinking. Always consider whether they need to be converted into the correct units. For example, times need to be in seconds rather than minutes.

Work, energy and potential difference

When an electrical force causes charge carriers to move, electrical energy is used and work is done. The amount of work depends on the charge and the potential difference (p.d.) through which the charge is moved. The work done is equal to the energy given to the charge carriers.

The p.d. between two points is equal to the work done in moving 1 C of charge from one point to the other.

Key points

* The SI unit of p.d. is the **volt** (V).
* The symbol for p.d. used by SQA is V.
* The relationship between charge, p.d. and work is $E_w = QV$.
* 1 volt = 1 joule per coulomb.
* When charge is moved in an electric field, the work done depends only on the start and finish points. It does not depend on the path followed by the charge.

Circuit symbols

You need to be able to draw and identify all the symbols in Table 1.1.

Table 1.1 Circuit symbols

Component	Symbol	Component	Symbol
ammeter	—Ⓐ—	voltmeter	—Ⓥ—
battery	⊣⊢⊣⊢	resistor	⊣▭⊢
fuse	⊏▭⊐	variable resistor	
switch		lamp	—⊗—
cell	⊣⊢	electrical supply	—o o—

Series circuits

Components connected one after another are connected in series. A circuit in which all the components are connected in this way is called a **series circuit**. In a series circuit there is one complete path of conductors. The series circuit in Figure 1.2 contains a battery, switch, ammeter, resistor and lamp.

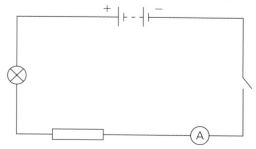

Figure 1.2 A series circuit

When you are using an ammeter to measure the current in an electrical component, the ammeter must **always** be connected **in series** with that component.

Parallel circuits

Components that are connected side by side are connected in parallel. A circuit in which components are connected in this way is called a **parallel circuit**. In a parallel circuit there is more than one complete path of conductors. The parallel paths are called **branches**.

Figure 1.3 shows a battery connected to two resistors and a voltmeter in parallel.

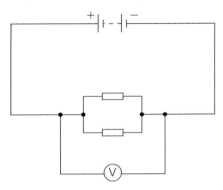

Figure 1.3 A parallel circuit

When you are using a voltmeter to measure the voltage across an electrical component, the voltmeter must **always** be connected **in parallel** with that component.

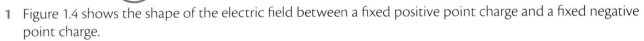

Questions ?

1 Figure 1.4 shows the shape of the electric field between a fixed positive point charge and a fixed negative point charge.

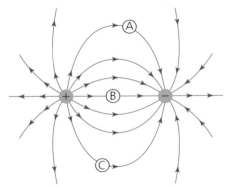

Figure 1.4

a) A stationary electron is located on a field line at point A.
   i)   Describe the path along which the electron will move.
   ii) State the final position of the electron.
b) A stationary neutron is located on a field line at point B.
   i)   State the final position of the neutron.
   ii) Justify your answer.

c) A stationary proton is located on a field line at point C.
 i) Describe the path along which the proton will move.
 ii) State the final position of the proton.

2 Figure 1.5 shows the shape of the electric field between two fixed negative point charges.

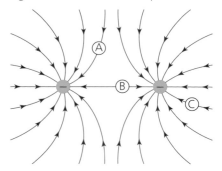

Figure 1.5

a) A stationary proton is located on a field line at point A.
 i) Describe the path along which the proton will move.
 ii) State the final position of the proton.
b) A stationary neutron is located at point B on the straight line between the centres of the two point charges.
 i) State the final position of the neutron.
 ii) Justify your answer.
c) A stationary electron is located on a field line at point C.
 i) State the direction of the electrical force acting on the electron.
 ii) Justify your answer.
 iii) Give a reason why it is not possible to state the final position of the electron.

3 State the two conditions that are necessary for charge to flow in a circuit.
4 Explain why both conductors and insulators are necessary for the operation of electrical circuits.
5 Name the charge carriers in each of the following:
 a) domestic electrical circuits
 b) electrolysis of copper sulphate solution
 c) semiconductor components in electronic circuits.
6 Pure water is an electrical conductor even though water particles are neutral molecules. Explain.
7 During a flash of lightning 600 C of charge is transferred in 0·15 s. Calculate the average current.
8 The work done in moving 1 C of charge around an electric circuit is 15 J. State the potential difference between the terminals of the supply.
9 Calculate the work required to move 40 mC through a potential difference of 1·8 kV.

10 Points A, B, C and D are in an electric field.

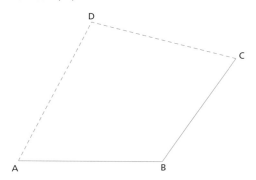

A charge of 0·30 C is moved from point A to point B and then to point C. The total work done is 1·2 J.

a) Calculate the potential difference between A and C.

b) A second charge of 0·30 C is moved from point A to point D and then to point C. State the total work done in moving this charge. Justify your answer.

11 Electrons are accelerated through a potential difference of 2·30 kV. The charge of an electron is $1·6 \times 10^{-19}$ C. The mass of an electron is $9·11 \times 10^{-31}$ kg.

a) Calculate the kinetic energy gained by the electrons.

b) Hence calculate the final speed of the electrons.

c) State any assumption you have made in your calculation.

12 Draw a series circuit containing a battery, switch, lamp, variable resistor and ammeter, connected so that it can be used to measure the current in the lamp.

13 Draw a circuit with a battery and switch in series, two lamps in parallel and a voltmeter connected so that it can be used to measure the voltage across one of the lamps.

14 Copy and complete the following table.

Component	Symbol	What it does
cell		
		Measures current in electrical circuits
resistor		
	─o o─	
		Turns current on and off
fuse		
		Source of light
	─(V)─	
variable resistor		
	─┤├-┤├─	Source of electrical energy

7

1.2
Electrical circuits

What you should know

For N5 Physics you need to be able to:

★ understand and use the terms resistance and resistor
★ understand and apply rules for current and voltage in series circuits
★ understand and apply rules for current and voltage in parallel circuits
★ understand and use the relationship $V = IR$
★ solve problems on electrical circuits.

Resistance

Resistance is a measure of the opposition to flow of charge in electrical circuits. Increasing resistance in a circuit causes current to decrease.

Key points

✳ The SI unit of resistance is the **ohm** (Ω – Greek letter omega).
✳ The symbol for resistance used by SQA is R.
✳ The relationship between current, voltage and resistance is $V = IR$.

A resistor is an electrical component with a resistance that remains approximately constant for different currents. Resistors often heat up during use and this usually causes the resistance of the resistor to increase slightly.

Example

An electric torch has two 1·5 V cells in series. The current in the bulb of the torch is 0·45 A. Calculate the resistance of the torch bulb.

$R = ?$ $V = IR$

$I = 0·45\,A$ $\Rightarrow 3·0 = 0·45 \times R$

$V = 1·5 + 1·5 = 3·0\,V$ $\Rightarrow R = 6·67 = 6·7\,\Omega$

Current and voltage in series circuits

Key points

In a series circuit:

* the current is the **same** at all points in the circuit
* the **sum** of the p.d.s across the individual components is equal to the voltage of the supply.

(*Learn these rules – and practise using them!*)

Example

In the circuit in Figure 1.6 the current in the 5·0 Ω resistor is 0·50 A.

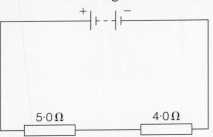

Figure 1.6

a) State the current in the 4·0 Ω resistor.
Current in the 4·0 Ω resistor = 0·50 A
(*Series circuit so current is the same at all points.*)

b) Calculate the battery voltage.
First, calculate the potential difference across each resistor.

$V = ?$
$I = 0.50\,A$
$R = 5.0\,\Omega$

$$V = IR$$
$$\Rightarrow V_{5\Omega} = 0.50 \times 5.0$$
$$= 2.5\,V$$

$I = 0.50\,A$
$R = 4.0\,\Omega$

$$\Rightarrow V_{4\Omega} = 0.50 \times 4.0$$
$$= 2.0\,V$$

Series circuit, so battery voltage = sum of potential differences across components

$$= 2.5 + 2.0 = 4.5\,V$$

Hints & tips

When cells with the same polarity (i.e. with the positive and negative terminals the same way round) are connected in series, the total voltage is equal to the sum of the p.d.s of the cells.

When cells with opposite polarity (i.e. with the positive and negative terminals opposite to each other) are connected in series, the total voltage is reduced and may be zero.

The same rules apply to batteries connected in series.

Current and voltage in parallel circuits

Key points

In a simple parallel circuit the parallel components are connected directly to the supply. In this kind of circuit:

* the sum of the currents in the parallel branches is equal to the current in the supply
* the potential difference across each parallel branch is equal to the supply voltage.

(*Learn these rules – and practise using them!*)

Hints & tips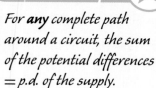

*For **any** complete path around a circuit, the sum of the potential differences = p.d. of the supply.*

Example

In the circuit in Figure 1.7 the potential difference across the $3.0\,\Omega$ resistor is $9.0\,V$.

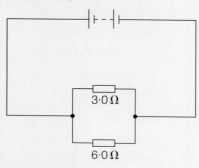

Figure 1.7

a) State the battery voltage.

   Battery voltage $= 9.0\,V$ (*same as p.d. across the $3.0\,\Omega$ resistor*).

b) Calculate the current in the $6.0\,\Omega$ resistor.

   p.d. across $6.0\,\Omega$ resistor $=$ battery voltage $= 9.0\,V$

   $I = ?$ $V = IR$

   $V = 9.0\,V$ $\Rightarrow 9.0 = I \times 6.0$

   $R = 6.0\,\Omega$ $\Rightarrow I = 1.5\,A$

More complex circuits

Many electrical circuits have a combination of series and parallel parts. In these circuits the above rules are modified slightly.

Key points ❗

For any series part of a circuit:

✳ the current is the **same** in all components connected in series
✳ the p.d. across the series part is equal to the **sum** of the potential differences across the individual components.

For any parallel part of a circuit:

✳ the total current in the parallel part equals the **sum of the currents** in the parallel branches
✳ the p.d. across each parallel branch is the **same**.

(*Learn these rules – and practise using them!*)

Example 🚩

In the circuit in Figure 1.8 the current in one of the 20 Ω resistors is 0·24 A.

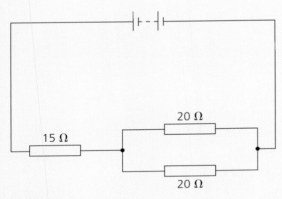

Figure 1.8

a) State the current in the other 20 Ω resistor.
  Current in the other 20 Ω resistor = 0·24 A (*p.d. across both resistors is the same.*)
b) State the current in the 15 Ω resistor.
  Current in the 15 Ω resistor = 0·48 A (*Sum of currents in parallel branches.*)
c) Calculate the battery voltage.
  p.d. across either parallel branch = IR
  $$= 0·24 \times 20 = 4·8\,V$$
  p.d. across 15 Ω resistor = $0·48 \times 15 = 7·2\,V$
  Battery voltage = p.d. across 15 Ω resistor + p.d. across parallel branches
  $$= 7·2 + 4·8 = 12\,V$$

There are questions on other more complex circuits in the Questions box below and in Topic 1.3. You can use the solutions to learn how to tackle questions like these.

Questions

1   A student connects two resistors in series with a 12 V power supply as shown in Figure 1.9.

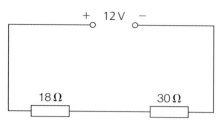

Figure 1.9

The current in the 18 Ω resistor is 0·25 A.
a) Calculate the potential difference across the 30 Ω resistor.
b) Calculate the total resistance of the circuit.

2   Two resistors are connected in parallel to a 6·0 V battery as shown in Figure 1.10.

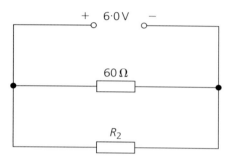

Figure 1.10

The current in the battery is 0·30 A.
a) Calculate the current in the 60 Ω resistor.
b) Calculate the resistance of resistor R_2.

3   Three resistors are connected to a 20 V supply as shown in Figure 1.11.
    The current in the 2·0 Ω resistor is 2·0 A.

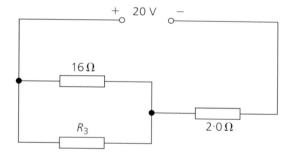

Figure 1.11

a) State the current in the battery. Explain your answer.
b) Calculate the p.d. across the 16 Ω resistor.
c) Calculate the resistance of resistor R_3.

1.3
Resistors

> ### What you should know
>
> For N5 Physics you need to be able to:
> * ★ understand and use the term potential divider
> * ★ understand and use the relationships $R_T = R_1 + R_2 + ...$,
> $$\frac{1}{R_T} = \frac{1}{R_1} + \frac{1}{R_2} + ...$$
> * ★ solve problems on resistors in circuits
> * ★ describe an experiment to verify Ohm's Law.

Measuring resistance

The circuit in Figure 1.12 is used to measure the resistance of resistor R.

The switch is closed and the meter readings are noted. The variable resistor is adjusted and the meter readings are again noted. This procedure is repeated until at least five pairs of meter readings, **over a sufficiently wide range of current values**, have been obtained.

The experimental measurements may be used to find the resistance by calculation **or** by using a graph.

By calculation: $V = IR \Rightarrow R = \dfrac{V}{I}$. For each pair of meter readings calculate the value of $\dfrac{V}{I}$ and then calculate the average value.

By graph: Plot a graph of the voltage readings on the y-axis against the current readings on the x-axis. Draw a best-fitting straight line. The resistance is equal to the gradient of this line.

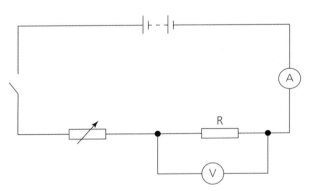

Figure 1.12 Measuring resistance

> ### Hints & tips ★
>
> *The above method is used to verify Ohm's Law.*
>
> *Showing that the resistance of an electrical component is constant over a wide range of current values verifies Ohm's Law for that component. We say that the component 'obeys Ohm's Law'.*
>
> *Many electrical components do not obey Ohm's Law.*

Example

Two students obtain the following readings using the above experimental method.

Voltmeter reading (V)	2·0	2·5	3·0	3·5	4·0	4·5
Ammeter reading (A)	0·17	0·21	0·26	0·30	0·34	0·39

a) Use these readings to obtain a value for the resistance of the resistor.
First calculate $\frac{V}{I}$ for each pair of readings.

Resistance (Ω)	11·76	11·90	11·54	11·67	11·76	11·54

Average value of resistance = (11·76 + 11·90 + 11·54 + 11·67 + 11·76 + 11·54) ÷ 6

$$= 11·695\,\Omega = 11·7\,\Omega$$

b) Suggest one way of improving the accuracy of this experiment.
Take more readings over a wider range **or** use a voltmeter with a much higher resistance.

c) The actual resistance of the resistor is 12 Ω. Suggest a reason why the measured value is lower than the true resistance.
The ammeter is measuring current in parallel branches (resistor + voltmeter), meaning the measured values of current are too big and so the calculated value of resistance is too small.

Combining resistors

For two or more resistors connected in **series**, the total resistance R_T is found using the relationship $R_T = R_1 + R_2 + \ldots$

Key points

* When another resistor is connected in series the total resistance increases.
* There is a single path so charge carriers have to go through an extra resistor.

For two or more resistors connected in **parallel**, total resistance R_T is found using the relationship $\frac{1}{R_T} = \frac{1}{R_1} + \frac{1}{R_2} + \ldots$

Key points

* When another resistor is connected in parallel the total resistance decreases.
* There is an extra path for charge carriers to follow so more charge flows.

Examples

1 Calculate the total resistance of the combination of resistors in Figure 1.13.

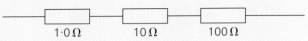

Figure 1.13

$R_1 = 1{\cdot}0\,\Omega$ $R_T = R_1 + R_2 + R_3$

$R_2 = 10\,\Omega$ $\Rightarrow R_T = 1{\cdot}0 + 10 + 100$

$R_3 = 100\,\Omega$ $= 111\,\Omega$

$R_T = ?$

2 Calculate the total resistance of the combination of resistors in Figure 1.14.

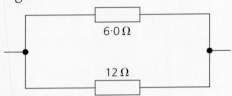

Figure 1.14

$R_1 = 6{\cdot}0\,\Omega$ $\dfrac{1}{R_T} = \dfrac{1}{R_1} + \dfrac{1}{R_2}$

$R_2 = 12\,\Omega$ $\Rightarrow \dfrac{1}{R_T} = \dfrac{1}{6{\cdot}0} + \dfrac{1}{12} = 0{\cdot}25$

$R_T = ?$ $\Rightarrow R_T = 4{\cdot}0\,\Omega$

Hints & tips

*When using the relationship for resistance in parallel, many people forget to **invert after** **adding** $\frac{1}{R_1} + \frac{1}{R_2}$ – do not let this happen to you.*

Potential divider

A potential divider consists of two resistors in series with a battery, as shown in Figure 1.15.

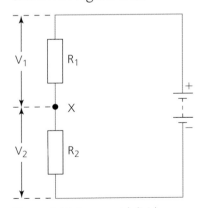

Figure 1.15 A potential divider

By choosing the resistors carefully we can control the voltage at point X. Potential dividers are used for controlling voltage in many electronic circuits.

The resistors are connected in series, so current in R_1 = current in R_2 = I.

$\Rightarrow V_1$ (voltage across R_1) = IR_1 **and** V_2 (voltage across R_2) = IR_2

$\Rightarrow \dfrac{V_1}{V_2} = \dfrac{IR_1}{IR_2} = \dfrac{R_1}{R_2}$

This means that the voltages across the resistors are in the ratio of their resistances.

Key points

Useful relationships:

$$\text{p.d. across } R_1 = \left(\frac{R_1}{R_1 + R_2}\right) V_s$$

$$\text{p.d. across } R_2 = \left(\frac{R_2}{R_1 + R_2}\right) V_s$$

where V_s is the voltage of the supply (in this case the battery voltage).

Example

A potential divider is made by connecting two resistors in series with a 9·0 V battery as shown in Figure 1.16.

Calculate the voltage across the 600 Ω resistor.

$R_1 = 300\,\Omega$

$R_2 = 600\,\Omega$

$V_s = 9{\cdot}0\,V$

$V_2 = ?$

$$V_2 = \left(\frac{R_2}{R_1 + R_2}\right) V_s$$

$$\Rightarrow V_2 = \left(\frac{600}{300 + 600}\right) \times 9{\cdot}0$$

$$= 6{\cdot}0\,V$$

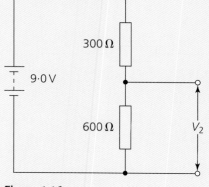

Figure 1.16

Questions

1 a) Use the experimental results from the Example box in the *Measuring resistance* section on page 14 to plot a graph of voltmeter reading (*y*-axis) against ammeter reading (*x*-axis). Use your graph to calculate the resistance of the resistor.
   b) The graph must pass through the origin, that is, point (0, 0). Explain why.
2 A student is carrying out an experiment with ten 10 Ω resistors. The student connects the resistors first in series and then in parallel.
   a) Calculate the total resistance of the ten resistors when they are connected in series.
   b) Calculate the total resistance of the ten resistors when they are connected in parallel.
3 A circuit is set up as shown in Figure 1.17.
   a) Calculate the total circuit resistance.
   b) Calculate the potential difference across the 8·0 Ω resistor.

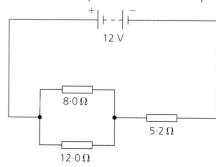

Figure 1.17

4 A circuit is set up as shown in Figure 1.18.

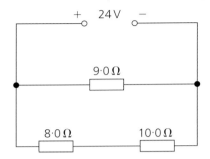

Figure 1.18

a) Calculate the total circuit resistance.

b) Calculate the potential difference across the 8·0 Ω resistor.

5 A student sets up the potential divider circuit in Figure 1.19.

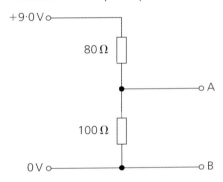

Figure 1.19

a) Calculate the voltage between points A and B.

b) State the p.d. across the 80 Ω resistor. Justify your answer.

6 In the circuit in Figure 1.20 the voltage of the battery is 6·0 V.

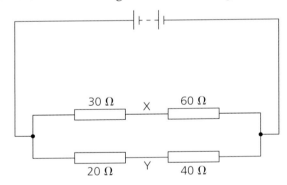

Figure 1.20

a) Calculate the total resistance of the circuit.

b) Calculate the circuit current.

c) Calculate the current in the 60 Ω resistor.

d) Calculate the current in the 20 Ω resistor.

e) Calculate the potential difference between X and Y.

1.4
Electrical energy

What you should know

For N5 Physics you need to be able to:

★ state and use the principle of conservation of energy

★ state that energy can change from one form to one or more other forms of energy

★ understand and use the relationships $E = Pt$, $P = VI$, $P = I^2R$ and $P = \dfrac{V^2}{R}$

★ carry out calculations on electrical power, electrical energy and time

★ identify energy changes in lamps and heaters

★ understand and use the terms power, d.c., a.c. and frequency

★ state that the quoted voltage of an a.c. supply is less than its peak voltage

★ state that a d.c. supply and an a.c. supply of the same voltage deliver the same power

★ identify a source as a.c. or d.c. based on an oscilloscope trace or image from data logging software

★ select an appropriate fuse given the power of an electrical appliance.

The principle of conservation of energy is an important one in Physics. It is used to describe and explain phenomena when dealing with almost any event. It includes mechanical, electrical, radioactive, electronic and thermal situations.

The key theme is that the total numerical value for the 'energy' before an event will be the same as the total numerical value for the 'energy' after the event.

A simple example is that of a light bulb (see Figure 1.21).

1000 J   100 light

900 heat

Figure 1.21 A light bulb   **Figure 1.22** Energy transformed by a light bulb

When switched on for a short time 1000 J of electrical energy is transformed to light and heat (Figure 1.22).

If 100 J is transformed to light then 900 J is transformed to heat. The number 1000 is the same before and after the event. It has been conserved.

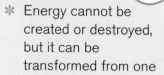

Key point !

❋ Energy cannot be created or destroyed, but it can be transformed from one form into another.

Energy and power

When there is current in an electrical component an energy change takes place. For example, in a lamp electrical energy is changed to heat and light. In the resistance wire of a heater electrical energy is changed to heat. In some heaters light is also given out.

The quantity of electrical energy changed to other forms **each second** is called **power**. Power is a scalar. The relationship between electrical energy, electrical power and time is $E = Pt$.

Key points

* The SI unit of power is the **watt** (W).
* The symbol for power used by SQA is P.
* The following three relationships are used for calculating electrical power:

$$P = VI = I^2R = \frac{V^2}{R}$$

Make sure you understand and know how to use all of them.

Example

The circuit in Figure 1.23 shows two resistors in series with a battery.

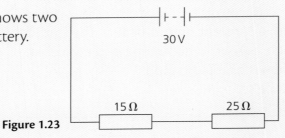

Figure 1.23

Calculate the energy transformed each second in the 25 Ω resistor.

(*First calculate current in the circuit.*)

Total resistance of circuit $= 15 + 25 = 40\,\Omega$

$R = 40\,\Omega$ $V = IR$

$V = 30\,V$ $\Rightarrow 30 = I \times 40$

$I = ?$ $\Rightarrow I = 0.75\,A$

(*Now use $P = I^2R$ to calculate the power.*)

$I = 0.75\,A$ $P = I^2R$

$R = 25\,\Omega$ $\Rightarrow P = (0.75)^2 \times 25$

$P = ?$ $= 14.06 = 14\,W$

d.c. and a.c. energy supplies

The abbreviation 'd.c.' is short for **direct current**. A d.c. supply causes charge to move in one direction. A supply with a constant voltage, like a battery, is a d.c. supply. This does **not** imply **constant** voltage, however – many d.c. supplies have variable voltages.

The abbreviation 'a.c.' is short for **alternating current**. An a.c. supply causes charge to move one way then the other. Mains electricity is an a.c. supply.

The quoted value of the mains is **230 V** and its peak voltage is 325 V. The **frequency** of the mains is 50 Hz – there are 50 complete cycles of positive and negative voltage each second.

The quoted value of voltage of an a.c. supply is always less than its peak value.

A graph of mains a.c. voltage against time has the shape shown in Figure 1.24.

An a.c. supply and a d.c. supply of equal quoted voltage supply the same power.

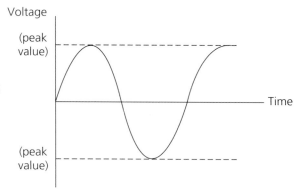

Figure 1.24 How an a.c. voltage varies with time

Example

A student has a d.c. supply with a constant output of 12 V. The student also has an a.c. supply of variable output. What setting of the variable a.c. supply delivers the same power as the d.c. supply?
The variable a.c. supply should be set to 12 V.

Fuses and safety

It is important that every piece of electrical equipment in your home has an appropriate fuse in the plug that connects the appliance to the socket. A fuse is a safety device that stops current from rising to a dangerous level. Fuses for mains plugs consist of a ceramic cylinder containing a thin wire connected to a metal cap on each end. If the current rises above a certain level, the thin wire melts and this stops the current. When you buy a new appliance the manufacturer will supply the correct fuse in the plug.

The most common fuses used in the UK have ratings of 3 A and 13 A. If you have to select an appropriate fuse for an appliance, you need to know its power rating. Appliances of power up to 720 W *usually* need a 3 A fuse. Appliances of power more than 720 W *usually* need a 13 A fuse.

If you have any doubt about the right fuse for any piece of electrical equipment, contact the manufacturer.

Example

Select the correct fuse for the following appliances.

a) Electric fire of power 3000 W

Appropriate fuse is 13 A.

b) Slow cooker of power 200 W

Appropriate fuse is 3 A.

Questions

1 In an electric fire where does the energy change take place?
2 State the energy change in a filament lamp.
3 **a)** Starting from $P = IV$ show that $P = I^2R$.

   b) Starting from $P = IV$ show that $P = \dfrac{V^2}{R}$.

4 A label on an electric kettle contains the following data: Power 3·0 kW; Voltage 230 V.

   a) Calculate the current in the kettle when it is on.

   b) How much electrical energy does the kettle use in 5 minutes?

5 A mains operated electric toaster has four identical elements connected in parallel, as shown in Figure 1.25.

   The power rating of the toaster is 1·84 kW.

   Calculate the resistance of each element.

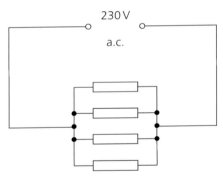

Figure 1.25

6 Select an appropriate fuse for each of the following appliances:

   a) Electric iron of power 1300 W

   b) Television of power 350 W

   c) Kettle of power 2500 W

   d) DVD player of power 35 W

7 In a kitchen a man is ironing shirts while his son is just about to switch on the kettle to make tea. The iron and the kettle are connected to the same 13 A extension lead.

   a) Assuming the kettle and iron have the same ratings as those in question 6, what should happen when the son switches on the kettle?

   b) Justify your answer by calculation.

1.5
Semiconductor diodes

What you should know

For N5 Physics you need to be able to:
* ★ describe the structure of semiconductor materials
* ★ draw and identify the symbol for a diode
* ★ describe the operation of a p-n junction.

Diodes are made from solid semiconductor materials. Semiconductors are the basis of **all** modern electronic equipment.

Semiconductors

Semiconductors are neither conductors nor insulators. The resistance of semiconductors is much lower than the resistance of insulators and much higher than the resistance of conductors.

Silicon is a semiconductor that was widely used in the early development of electronics. In the Periodic Table of elements, silicon is on the boundary between metals and non-metals **and** it is in the same column as carbon. Like carbon it has four electrons in its outer shell. Atoms like these form strong and stable crystals (we call carbon crystals diamonds).

n-type and p-type

Like diamonds, pure silicon crystals are insulators. To make them into semiconductors we must scatter a small number of other atoms throughout the crystals. This is called **seeding**.

To make p-type and n-type silicon the other atoms must be about the same size as the silicon atoms. They must also have **three** or **five** electrons in their outer shells. Aluminium and phosphorus atoms are suitable to use.

Figure 1.26 represents the structure of pure silicon, p-type silicon and n-type silicon.

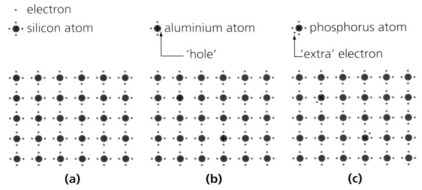

Figure 1.26 (a) Pure silicon; **(b)** p-type silicon; **(c)** n-type silicon (The actual crystal structure is 3D. It is tetrahedral – like a pyramid with a triangular base and all four faces the same size and shape)

Every silicon atom has four outer electrons and four adjacent atoms. In pure silicon all of these electrons form covalent bonds with adjacent atoms. There are no free electrons so the crystal is an insulator.

To make p-type silicon, crystals are seeded with aluminium atoms, which have only **three** outer electrons. Each aluminium atom is surrounded by **four** silicon atoms. Electrons from three of these silicon atoms form covalent bonds. An electron from the fourth silicon atom cannot form a bond, which leaves a space in the crystal structure. We call these spaces **holes**.

When a d.c. voltage is applied to p-type silicon, the holes move towards the negative terminal. The charge carriers in p-type silicon are holes.

To make n-type silicon, crystals are seeded with phosphorus atoms, which have **five** outer electrons. Each phosphorus atom is surrounded by **four** silicon atoms. Electrons from all of the silicon atoms form covalent bonds. This leaves one 'free' electron.

When a d.c. supply is connected to the n-type silicon, the free electrons move towards the positive terminal. The charge carriers in n-type silicon are electrons.

Both p-type and n-type silicon are electrically neutral, as the total number of electrons is equal to the total number of protons.

Key points

* In p-type silicon the charge carriers are holes, which move towards the negative terminal.
* In n-type silicon the charge carriers are free electrons, which move towards the positive terminal.

Example

Look up a Periodic Table of the elements. Find the locations of aluminium, silicon and phosphorus.

a) From the positions of these elements, what can you conclude about the sizes of their atoms?
   Aluminium and phosphorus are in the same row as silicon and are either side of silicon. The atoms of aluminium and phosphorus will therefore be about the same size as silicon atoms.

b) Why is the size of aluminium and phosphorus atoms important?
   Atoms that are about the same size as silicon can fit well into the crystal structure. Atoms that are much smaller or much bigger than silicon atoms would distort the crystal and so seriously weaken its strength and stability.

p-n junction

When we join together a piece of p-type and a piece of n-type silicon, we make a p-n junction. The normal thermal vibration of atoms causes some free electrons in the n-type to cross the junction into the p-type. These electrons form bonds where there are holes – effectively filling holes in the p-type.

The area near the junction is called the **depletion layer**. There are no charge carriers in the depletion layer. Free electrons have left the n-type and the holes in the p-type have been filled. Just like in pure silicon all the crystal bonds are complete, so this layer is an insulator.

Figure 1.27 shows the two ways of connecting a d.c. supply to a p-n junction.

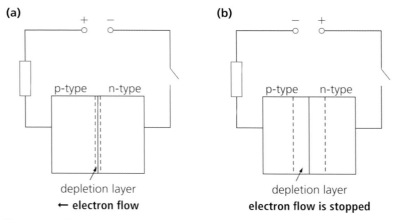

Figure 1.27 Connecting a p-n junction to a d.c. supply

When the switch is closed the electric field pushes electrons away from the negative terminal and towards the positive terminal.

Figure 1.27(a): Electrons are pushed into the n-type and pulled out of the p-type. Electrons from the supply replace free electrons in the n-type at the same time as electrons are pulled out of holes in the p-type. The depletion layer contracts and disappears. The p-n junction conducts.

Figure 1.27(b): Electrons are pushed into the p-type, filling more holes, at the same time as free electrons are pulled out of the n-type. The depletion layer expands and the depth of insulator increases. The p-n junction does not conduct.

Diodes

A diode is a device that allows electrons to flow in one direction only. It behaves like a conductor in one direction and an insulator in the other. A p-n junction is a diode.

Diodes are used when an a.c. source like mains electricity is the supply for electronic equipment or when an electronic circuit needs to prevent a flow of electrons in a particular direction.

Key points

The symbol for a diode used by SQA is:

Note the shape of an arrow in this symbol. Electrons flow in the **opposite direction** to the arrow – from **right** to **left** in this case.

Hints & tips

The arrow in the symbol for a diode points from p-type to n-type.
To make a diode conduct, connect the p-type to positive and the n-type to negative.

Example

The diode in the circuit in Figure 1.28 is at its operating current of 24 mA.

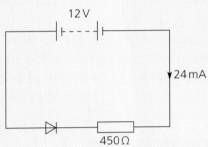

a) Calculate the p.d. across the diode.
First calculate the p.d. across the resistor:

$I = 24\,mA = 2.4 \times 10^{-2}\,A$ $V = IR$
$R = 450\,\Omega$ $\Rightarrow V = 2.4 \times 10^{-2} \times 450$
$V_{battery} = 12\,V$ $= 10.8\,V$

$$\text{p.d. across diode} = V_{battery} - V_R$$
$$= 12 - 10.8 = 1.2\,V$$

b) What is the purpose of the resistor in this circuit?
The purpose of the resistor is to protect the diode.

Figure 1.28

Michael Faraday observed a semiconductor effect in 1833 but it was not until the transistor was invented in the early 1950s that the potential of solid-state electronics began to be exploited.

Originally physicists used naturally occurring semiconductor materials like silicon or germanium. Now a wide variety of synthetic materials are used. There are even organic semiconductors – this is an area of science that uses Physics, Chemistry and Biology.

Modern semiconductor materials have increased reliability and have reduced the size, weight and cost of electronic equipment.

Questions

1 Germanium is a semiconductor. Look up a Periodic Table of elements to find the position of germanium.
  a) Suggest an element suitable for seeding germanium crystals to produce p-type germanium. Justify your suggestion.
  b) Suggest an element suitable for seeding germanium crystals to produce n-type germanium. Justify your suggestion.
2 Which, if any, parts of a p-n junction are electrically charged?
3 State two benefits of modern electronics.

1.6
Capacitors

What you should know

For N5 Physics you need to be able to:

★ describe the structure of a capacitor
★ draw and identify the circuit symbol for a capacitor
★ explain why work must be done to charge a capacitor
★ describe and explain the behaviour of capacitors in d.c. and a.c. circuits
★ describe and explain possible uses of capacitors.

A capacitor consists of two metal plates separated by insulating material. When a d.c. voltage is connected to a capacitor, one plate becomes positively charged and the other becomes negatively charged. Electrical energy is stored in the electric field between the plates.

Key points

The symbol for a capacitor used by SQA is:

Capacitors are used to provide energy quickly, for example for the flash of a camera. They are also used to reduce variation in the voltage of d.c. electrical supplies.

Capacitors in d.c. circuits

An initially uncharged capacitor is connected in the circuit shown in Figure 1.29.

When the switch is closed the ammeter reading starts high and gradually falls to zero. The voltmeter reading starts at zero and gradually increases to the supply voltage. The size of the initial current is determined by the resistance of the resistor and the supply voltage.

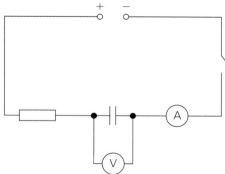

Figure 1.29

Example

A capacitor is connected in series with a switch, a 9·0 V battery and a 45 Ω resistor (Figure 1.30). The switch is closed.

a) Calculate the initial current.

The capacitor has no effect on the initial current.

$$V_{battery} = 9·0\,V \qquad V = IR$$
$$R = 45\,\Omega \qquad\qquad \Rightarrow 9·0 = I \times 45$$
$$\Rightarrow I = \frac{9·0}{45}$$
$$= 0·20\,A$$

b) State the final current.

The final current = 0 A.

c) State the size of the final p.d. across the capacitor. Give a reason for your answer.

Final p.d. across the capacitor = 9·0 V. It is numerically equal to the battery voltage.

d) Sketch the circuit diagram with the switch closed. On your diagram mark the positive and negative plates of the charged capacitor.

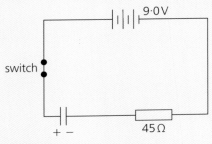

Figure 1.30

Key points

In a d.c. circuit the p.d. across a charged capacitor is *always* in the opposite direction to the supply voltage.

The graphs in Figure 1.31 show the shapes of the graphs of current in the capacitor and p.d. across the capacitor during charging.

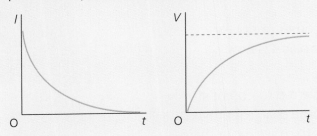

Figure 1.31

For N5 Physics you may be asked to draw and/or explain the shapes of graphs for charging and discharging capacitors. Question 1 at the end of this topic is about a capacitor discharging.

Capacitors in a.c. circuits

An initially uncharged capacitor is connected in the circuit shown in Figure 1.32.

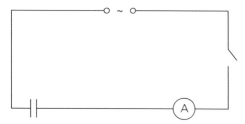

Figure 1.32

When the switch is closed a steady reading is observed on the a.c. ammeter.

In the first half cycle of the a.c. supply the capacitor begins to charge. In the next half cycle the capacitor discharges. This repeats during every cycle of the supply.

When the frequency of the a.c. supply is increased, the reading on the a.c. ammeter increases. The higher the a.c. frequency, the less time there is for the capacitor to charge.

Capacitors and energy

Consider two initially uncharged metal plates, A and B. Negative charge is moved from B to A.

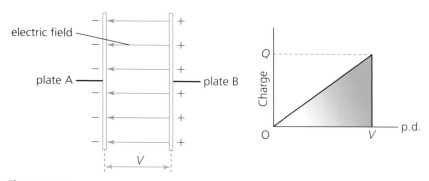

Figure 1.33

Plate A becomes negatively charged and plate B becomes positively charged. Negative charge being moved between the plates is now repelled by A and attracted by B.

Work done overcoming these electrical forces is converted to potential energy in the electric field between the plates. The more charge is moved from plate B to plate A the more difficult it becomes to move more charge from B to A and the greater the energy stored in the electric field becomes.

The graph in Figure 1.33 shows the relationship between the potential difference between the plates and the quantity of charge transferred from plate B to plate A.

Uses of capacitors

In N5 Physics you could be asked to describe and/or explain the following uses of capacitors.

Storing energy: When a capacitor is charged, energy is stored in the electric field between its plates. This energy can be used to provide a short, high-energy pulse of current. A good example is the flash unit of a camera. A capacitor is charged from the camera battery. When the shutter is pressed the capacitor discharges through a flash tube, giving a pulse of bright light.

Reducing voltage variation: A capacitor connected in parallel with the output of a d.c. power supply smooths the output – that is, it reduces the size of ripples in the output. The p.d. across the capacitor stabilises the output of the supply because time is needed to move charge onto and off of the plates of the capacitor.

Blocking d.c. while passing a.c.: A mixed signal is one which has both d.c. and a.c. components. In order to read information carried by the a.c. part of the signal, the d.c. part must be removed. A capacitor connected in series with a mixed input signal charges up to the p.d. of the d.c. part of the signal. This removes the d.c. part while the a.c. part is allowed to pass.

Key points!

The energy stored in a charged capacitor is equal to:

* the total work done in overcoming electrical forces during charging
* the area under the charge (*y*-axis) versus p.d. (*x*-axis) graph as shown in Figure 1.33.

The relationship between energy stored, charge transferred and fully charged voltage is:

$$E = \frac{1}{2}QV$$

Questions

1   An initially charged capacitor is connected in the circuit shown in Figure 1.34.

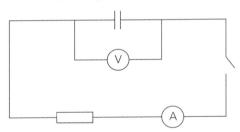

Figure 1.34

   The switch is closed.
   a) Describe what happens to the readings on the meters.
   b) Sketch graphs to show the
       i) current in the capacitor during discharging (values are not required)
       ii) p.d. across the capacitor during discharging (values are not required).
2   **a)** State the relationship between the magnitude of the excess charge on the plates of a capacitor and the potential difference between the plates.
   b) Why is the word 'excess' included in part a) of this question?

1.7

Electrical supplies for electronic circuits

The mains electricity is 230 V a.c. and many electrical and electronic circuits operate on much lower d.c. voltages. In order to use them we need to reduce the voltage **and** convert a.c. to d.c.

Reducing voltage

For many circuits, **step-down transformers** reduce the voltage from 230 V to values suitable for the electronic components.

All transformers are made by winding wire around iron or another 'magnetic' material. There is a **primary coil** of wire and a **secondary coil** of wire. There is no electrical connection between these coils – the connection is magnetic. In an electrical supply the primary coil is connected to the mains supply and the secondary coil is connected to the electronic circuit (Figure 1.35).

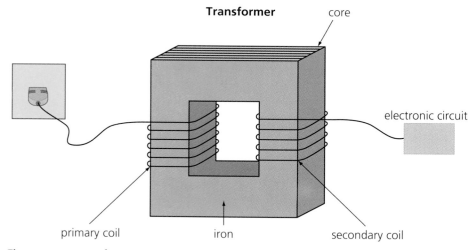

Figure 1.35 A transformer

When the a.c. voltage in the primary coil (V_p) is switched on it causes an a.c. current. This current causes a changing magnetic field. The iron core guides this magnetic field into the secondary coil. The changing magnetic field in the secondary coil produces an a.c. voltage across the secondary coil (V_s).

The size of the voltage in the secondary coil is controlled by the number of turns in the primary coil (n_p) and the number of turns in the secondary coil (n_s).

Key points

* Step-down transformers have fewer turns in the secondary coil than in the primary coil.
* The relationship between the primary and secondary voltages is $V_s = \dfrac{n_s}{n_p} V_p$.
* The frequency of V_s is equal to the mains frequency of 50 Hz.
* The secondary voltage may be changed to any value by changing the number of turns in the coils.

Example

An electronic engineer is designing a transformer for a power supply for a games console. The engineer wants to have 4600 turns in the primary coil and 9·0 V in the secondary coil. How many turns of wire should there be in the secondary coil?

$$V_p = 230\,V \qquad V_s = \frac{n_s}{n_p} V_p$$

$$V_s = 9{\cdot}0\,V \qquad \Rightarrow 9{\cdot}0 = \frac{n_s}{4600} \times 230$$

$$n_p = 4600 \qquad \Rightarrow n_s = 180 \text{ turns}$$

$$n_s = ?$$

Converting a.c. to d.c.

The process of converting a.c. to d.c. is called **rectification**. The part of the circuit that makes this change is called a **rectifier**. Diodes are the active components of rectifiers.

Figure 1.36 shows a rectifier with one diode. It also shows the shapes of the input and output voltages.

When terminal A is positive and terminal B is negative, the diode conducts and there is a p.d. across resistor R.

When terminal A is negative and terminal B is positive, the diode does not conduct and the p.d. across resistor R is zero.

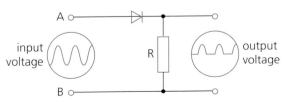

Figure 1.36 A rectifier with one diode, showing its input and output voltages

This type of circuit is called a **half wave** rectifier. Its output is not a constant voltage. For half of the time the output voltage is zero and for the rest of the time it is changing rapidly from zero to its peak value and back to zero. It is d.c., however, as the direction of electron flow never changes.

Figure 1.37 shows a better rectifier with four diodes. It also shows the shapes of the input and output voltages.

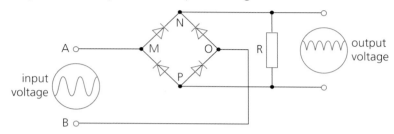

Figure 1.37 A rectifier with four diodes, showing its input and output voltages

When terminal A is positive and terminal B is negative, the diodes between MN and OP conduct and there is a p.d. across resistor R. The diodes between MP and NO do not conduct.

When terminal A is negative and terminal B is positive, the diodes between MP and NO conduct and there is a p.d. across resistor R. The diodes between MN and OP do not conduct.

This circuit is called a **full wave** rectifier. Its output is changing rapidly from zero to its peak value and back to zero all the time.

For electronic circuits we also need to reduce the variation in the voltage which is done using capacitors.

Smoothing

Figure 1.38 shows a capacitor connected across the output of a full wave rectifier. It also shows the shapes of the input and output voltages.

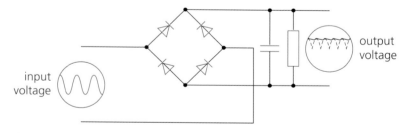

Figure 1.38 A capacitor used with a full wave rectifier

The dotted line in the trace on the right is the output of the rectifier. The continuous line is the smoothed output. There is now only a small ripple on the output voltage.

When the output voltage of the rectifier is increasing, the capacitor is charged to peak voltage. When the output voltage of the rectifier starts to decrease, the capacitor discharges slowly through the resistor. The voltage falls by a small amount. The next crest of the rectifier output recharges the capacitor and the process repeats.

Questions

1 Describe the difference between a.c. and d.c. electricity.

2 An electrical supply for an electronic circuit has an output voltage of 6·3 V. The number of turns on the secondary coil is 200. Calculate the number of turns on the primary coil.

3 Look again at Figure 1.37. Use the letters A, B, M, N, O, P and R to specify the path of electron flow when terminal A is:
   a) positive
   b) negative.

4 Electrical supplies for electronic circuits use full wave rectifiers rather than half wave rectifiers. Give a reason for this.

1.8
Input and output devices

All electronic systems have inputs and outputs.

Input devices

Input devices put energy into electronic circuits. All input devices change other form(s) of energy to **electrical energy**.

Table 1.2 includes details of common input devices.

Table 1.2 Common input devices

Input device	Symbol used by SQA	Energy change
microphone		sound → electrical energy
thermocouple		heat → electrical energy
solar cell		heat + light → electrical energy
thermistor		heat → electrical energy
LDR (light dependent resistor)		light → electrical energy

Input devices are used as **sensors** in electronic circuits. When a change around a sensor causes an electrical property of the sensor to change, this is used to activate an electronic system. For example, a **microphone** changes sound waves to electrical waves. Sound waves near a microphone cause electrical waves, which can be the input to switch on a voice-activated control system.

A **thermocouple** consists of two different metals joined together at one end and separated at the other. Usually the metals are insulated wires. When there is a temperature difference between the joined end and the separated end, a voltage is set up. When the temperature difference increases, the voltage increases. Thermocouples do not need batteries or any other source of energy. They get all the energy they need from their surroundings.

Solar cells convert light directly into electrical energy. When exposed to light, solar cells can generate and support an electric current. Solar cells do not need batteries or any other source of energy.

Thermistors are usually made from ceramic semiconductors. The resistance of most common thermistors decreases as temperature increases. This is unusual as the resistance of most materials increases as temperature increases.

Light dependent resistors (LDRs) are made from semiconductor materials with light-sensitive properties. The resistance of an LDR decreases as the brightness of light increases.

Example

Part of the circuit of an electronic system is shown in Figure 1.39.

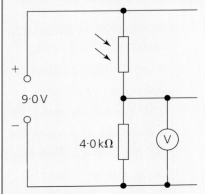

Figure 1.39

The reading on the voltmeter is 2·0 V.

a) Calculate the resistance of the LDR.

$$V_2 = 2{\cdot}0\,\text{V}$$

$$V_s = 9{\cdot}0\,\text{V}$$

$$R_2 = 4{\cdot}0\,\text{k}\Omega = 4000\,\Omega$$

$$R_1 = ?$$

$$V_2 = \left(\frac{R_2}{R_1 + R_2}\right) V_s$$

$$\Rightarrow 2{\cdot}0 = \left(\frac{4000}{R_1 + 4000}\right) \times 9{\cdot}0$$

$$\Rightarrow 2{\cdot}0 \times (R_1 + 4000) = 4000 \times 9{\cdot}0$$

$$\Rightarrow 2R_1 + 8000 = 36\,000$$

$$\Rightarrow R_1 = 14\,000\,\Omega\ (14\,\text{k}\Omega)$$

b) State one application of this type of circuit.
This circuit could be used as an automatic switch for lights, for example for switching on street lights when it gets dark.

Output devices

Output devices take energy out of electronic circuits. All output devices change **electrical energy** to one or more other forms of energy.

Table 1.3 includes details of common output devices.

Table 1.3 Common output devices

Output device	Symbol used by SQA	Energy change
LED (light emitting diode)		electrical energy → light
buzzer		electrical energy → sound
motor		electrical energy → kinetic energy
loudspeaker		electrical energy → sound

Output devices do things – they make things happen. For example, a voice-activated control system may switch on an electric motor to open a garage door. The electric motor is the output device – it makes the door open.

Like any other diode, an **LED** allows charge to flow in only one direction. When you draw a circuit with an LED you must make sure that it is connected correctly – see Figure 1.40.

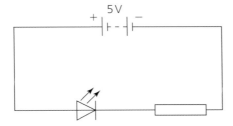

Figure 1.40 An LED connected correctly in a circuit

Hints & tips

Many Physics students lose marks by drawing LED symbols the wrong way round. Make sure you get this correct! The arrow in the symbol for an LED points from p-type to n-type. To make an LED conduct, connect the p-type to positive and the n-type to negative.

Study this diagram carefully! The electron flow is in the opposite direction to the arrow. The resistor protects the LED by making sure that the current is not too big.

LEDs are often used to show that electrical equipment is on.

Example

An LED, connected to a 5·0 V battery as shown in Figure 1.40, is operating correctly. The potential difference across the LED is 1·6 V when it is at its operating current of 20 mA. Calculate the resistance of the resistor.

$I = 20\,mA = 0.020\,A$ $V = IR$

$V = 5.0 - 1.6 = 3.4\,V$ $\Rightarrow 3.4 = 0.020 \times R$

$R = ?$ $\Rightarrow R = 170\,\Omega$

Questions ?

1   Name a suitable sensor for each of the following:
   a) switching heating off at a selected temperature
   b) setting off an automatic fire alarm
   c) measurement of temperature
   d) turning street lights off in the morning.
2   Give one application for each of the following:
   a) a motor
   b) an LED
   c) a buzzer
   d) a loudspeaker.
3   State the energy change in:
   a) an electric motor
   b) a thermistor
   c) a buzzer
   d) an LDR.
4   With regard to energy, what do all input devices have in common?
5   Part of the circuit diagram of an electronic system is shown in Figure 1.41. The resistance of component X is 2·5 kΩ.

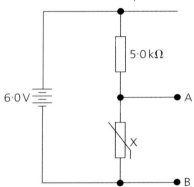

Figure 1.41

   a) Name component X.
   b) Calculate the p.d. between points A and B.
   c) State one possible application for this type of circuit.

1.9
Switches

> ### What you should know
>
> For N5 Physics you need to be able to:
> ★ draw and identify circuit symbols for an NPN transistor, an n-channel enhancement MOSFET and a relay
> ★ describe and explain how to switch on and switch off transistor switches
> ★ describe and explain the operation of relays.

Transistors are very fast electronic switches which are used for automatic control of electronic circuits.

NPN transistor

An NPN transistor has three terminals: the base, emitter and collector. It is made from two pieces of n-type semiconductor material and one piece of p-type semiconductor material. The structure and circuit symbol of an NPN transistor are shown in Figure 1.42.

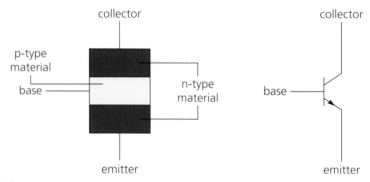

Figure 1.42

To switch on an NPN transistor, the base-emitter junction is forward-biased – that is, the p-type base is connected to positive and the n-type emitter is connected to negative. When the voltage at the base is at least +0·7 V above the voltage at the emitter, the transistor switches on and electrons flow through the transistor from emitter to collector. When the p.d. between the base and emitter falls below about 0·7 V the transistor switches off and the electron flow through the transistor stops.

n-channel enhancement MOSFET

A MOSFET is a **M**etal **O**xide **S**emiconductor **F**ield **E**ffect **T**ransistor.

The structure and circuit symbol of an n-channel enhancement MOSFET are shown in Figure 1.43. Note that the back electrode is connected to the source.

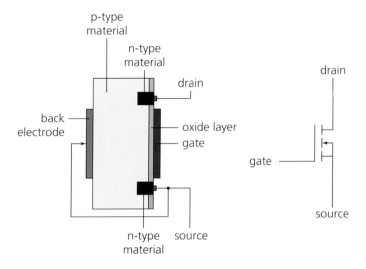

Figure 1.43

To switch on a MOSFET, a positive voltage of over 2 V is applied to the gate. This sets up an electric field between the gate and the back electrode. Electrons in the p-type material are attracted towards the gate. They gather in a layer near the gate and form a channel which enables electrons to flow from the source to the drain.

For N5 Physics you do not need to remember the values of base or gate voltage at which these transistors switch on. You do need to remember that they switch on at a set voltage.

Example

The circuit shown in Figure 1.44 is used to switch on the buzzer when the temperature rises above $-3\,°C$.

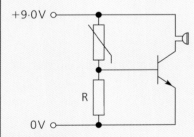

Figure 1.44

a) Explain how this circuit operates to switch on the buzzer.
(When the temperature is below $-3\,°C$ the transistor is off.)
As the temperature increases, the resistance of the thermistor falls. The voltage across resistor R therefore increases.
When the voltage across resistor R rises above 0·7 V, the transistor switches on and there is then a current in the buzzer.

b) Suggest one use for this circuit.
This circuit could sound an alarm when the temperature inside a freezer becomes too high.

Relay

A relay consists of an electromagnet (coils of wire wrapped around a core of iron or other magnetic material) and a switch made of iron or other magnetic material (Figure 1.45).

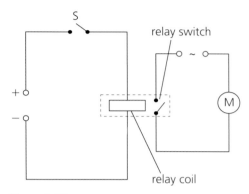

Figure 1.45

Switch S is closed. Current in the relay coil produces a magnetic field. The magnetic field attracts the relay switch, causing it to close. This completes the right-hand circuit and switches on the motor.

Relays in low voltage circuits are used to control high voltage and high power circuits. For example, relays are used to switch on heavy machinery in factories and motors in electric trains. The relays ensure that there is no danger of electrocution from the high voltages. Relays are also used where several circuits need to be switched on at the same time, for example street lighting.

Questions

1 Draw the circuit symbols for
   a) an NPN transistor
   b) a relay.
2 The circuit shown in Figure 1.46 is used to switch off a relay when the intensity of light incident on the LDR rises to a certain level. The relay then switches off a lamp.

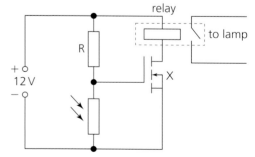

Figure 1.46

During the hours of darkness the resistance of the LDR is 5 kΩ and the voltage across the LDR is 6·0 V.
   a) Name component X.
   b) Calculate the resistance of resistor R.
   c) Describe how the circuit operates to switch off the relay.

Chapter 2: Properties of matter

Specific heat capacity and latent heat

What you should know

For N5 Physics you need to be able to:

★ understand and explain the difference between heat and temperature
★ know that different materials require different quantities of heat to raise their temperature by one degree Celsius (for the same mass)
★ know that temperature is a measure of the mean kinetic energy of the particles
★ know that heat flows from a hot object to a cold object
★ understand and use the relationship $E_h = cm\Delta T$
★ know that different materials require different quantities of heat to change state
★ understand and use the relationship $E_h = ml$.

Heat is a form of energy. This topic deals with the transfer of heat energy. Heat is often confused with temperature but they are **not** the same thing.

Key points

✳ Heat energy is measured in **joules**.
✳ It has the symbol E_h.

Heat can be transferred from a hot object to a cooler object. For example, in cold weather you feel 'cold' because heat is leaving your body, since you are hotter than the surrounding area. Temperature cannot be transferred.

Similarly, water in a kettle heats up because the element is at a higher temperature than the water, and so heat is transferred from the element to the water. Hot water in a cup left in a room cools down because heat is transferred from the water to the surroundings, which are at a lower temperature (Figure 2.1).

Figure 2.1 Heat energy is transferred to the surroundings

The principle of conservation of energy (see page 18) can be applied in all the situations discussed in this topic. No energy is ever lost, but it may be transformed to different types of energy.

When an object gains energy it increases its temperature. If we supply a piece of copper (say) with 500 J of energy, its temperature will increase.

Key points

* Change in temperature has the symbol ΔT.
* It is measured in degrees Celsius (°C).

If we transfer the same amount of heat energy to different materials (of the same mass) we find that their temperature increase is different. This difference is due to a property of each material called its **specific heat capacity**, c.

The change in temperature of an object depends on its mass, its specific heat capacity and the amount of heat energy transferred.

Key points

The relationship linking heat energy transferred and temperature change is:

$$E_h = cm\Delta T$$

where E_h = heat energy, measured in joules

c = specific heat capacity, measured in joules per kilogram per degree

ΔT = change in temperature, measured in °C (the symbol Δ, delta, means 'change in').

Example

24 000 J of energy is transferred to 2·50 kg of aluminium. Calculate its increase in temperature.

$E_h = 24\,000$ J

c for aluminium $= 910\,\mathrm{J\,kg^{-1}\,°C^{-1}}$

$m = 2.5$ kg

$E_h = cm\Delta T$

$24\,000 = 910 \times 2.5 \times \Delta T$

$\Delta T = \dfrac{24\,000}{910 \times 2.5} = 10.5\,°C$

Hints & tips

You're not expected to remember data such as that c for aluminium = $910\,\mathrm{J\,kg^{-1}\,°C^{-1}}$ – these will be given in the Data Sheet.

Example

50 000 J of energy is transferred to 3·0 kg of water at 22 °C. Calculate the temperature of the water after the energy has been transferred. Assume no energy losses.

$E_h = 50\,000$ J

c for water $= 4180\,\mathrm{J\,kg^{-1}\,°C^{-1}}$

$m = 3.0$ kg

$E_h = cm\Delta T$

$50\,000 = 4180 \times 3.0 \times \Delta T$

$\Delta T = \dfrac{50\,000}{4180 \times 3.0} = 4\,°C$

Final temperature of water $= 22 + 4 = 26\,°C$

Questions ?

Remember to look in the Data Sheet for values that are not given in the questions.

1 Calculate the energy required to increase the temperature of 15 kg of lead by 25 °C.
2 Calculate the energy required to increase the temperature of 150 g of water by 35 °C.
3 32 000 J increases the temperature of 2·0 kg of a material by 8·0 °C. Calculate a value for its specific heat capacity.
4 16 000 J increases the temperature of a mass of iron by 32 °C. Work out the mass of iron heated.
5 2·0 kg of water in a kettle is boiled and then allowed to cool. How much energy will it transfer to its surroundings if it cools to 25 °C?

The important principle in the transfer of heat energy is that energy is transferred from a hotter object to a cooler object. When you place your hand in a beaker of cold water, heat is transferred from your hand to the water. This transfer of energy causes the water to heat up and your hand to cool down.

Similarly, pouring cold milk into hot tea will cool the tea down to allow it to be drunk without burning the drinker's mouth. Or if we put cold potatoes into hot water, heat energy is transferred to the potatoes and the water cools down. In both of these cases energy is conserved. Heat energy is transferred from the hot tea to the colder milk, or from the hot water to the colder potatoes. This energy is gained by the milk and potatoes, which in turn increases their temperature.

If we put a metal at 50 °C into water at 50 °C, what happens? Essentially nothing. There is no difference in temperature so no heat energy will transfer from one object to the other.

Heat can be transferred by conduction, convection or radiation, but these processes are not discussed here.

How do hot objects differ from cold objects?

What happens to the energy we give to objects (bodies)? We know that if we transfer energy to an object then its temperature will rise, but what change in the object does this energy cause?

The energy will cause the particles in the object to gain energy and vibrate (move) more. The particles in a hot material vibrate more than those in a cold material. The heat energy given to the material is transformed into movement or kinetic energy in the particles of the material.

Temperature is often described in textbooks as a measure of the average kinetic energy of the particles. This seems complex but essentially just means that the particles in warmer substances are moving more quickly.

On the previous page, we considered what occurs when objects are heated or cooled, and how this affects any change in temperature.

In general, when we apply heat to an object its temperature will increase. There are some instances, however, when heat is applied to an object and its temperature remains the same. For example, when we boil water in a beaker, the water will reach 100°C and start to evaporate. Although we continue to heat the water, the temperature will not increase. In this case, the heat energy is being used to change the state of the material from liquid to gas. When a solid object gains enough energy it can change to a liquid. It can melt. This is also a change of state.

Latent heat is the energy required to change the state of a material. **Specific latent heat of fusion** refers to the energy required to change 1 kg of a solid to a liquid. (Fusion is an old word for melting.)

When a liquid gains enough energy it can change to a gas. This is called vaporisation. **Specific latent heat of vaporisation** refers to the energy required to change 1 kg of a liquid to a gas.

The relationship required for calculating the energy lost or gained during a change of state is:

$E_h = ml$

Where:

E_h is heat energy in joules (J).

m is mass, in kilograms (kg).

l is specific latent heat, in joules per kilogram $J\,kg^{-1}$

Example

A 5·5 kg mass of iron is melted in a furnace. Determine how much energy is required to melt the iron.

The latent heat of fusion of iron is $2·6 \times 10^5\,J\,kg^{-1}$

$\quad E_h = ml$

$\qquad = 5·5 \times 2·6 \times 10^5$

$\qquad = 1·4 \times 10^6\,J$

The molten iron continues to heat up and eventually evaporates. Calculate the energy required for the iron to change from a liquid to a gas.

Latent heat of vaporisation of iron is $61·2 \times 10^5\,J$

$\quad E_h = ml$

$\qquad = 5·5 \times 61·2 \times 10^5$

$\qquad = 3·4 \times 10^7\,J$

Questions

1. Calculate the energy required to melt 0·45 kg of ice at 0°C? (Latent heat of fusion of ice is $3·34 \times 10^5\,J\,kg^{-1}$.)
2. 240 000 J of energy was used to melt 1·3 kg of copper. Assuming all the energy was used only to melt the copper, calculate the value this gives for the specific latent heat of fusion for copper.

2.2
Gas laws and the kinetic model

What you should know

For N5 Physics you need to be able to:
* ★ state that pressure is the force per unit area
* ★ use the relationship $P = \dfrac{F}{A}$ appropriately
* ★ explain situations using the motion of the particles
* ★ convert temperature from Kelvin to Celsius and vice versa
* ★ perform calculations using the relationships between pressure, volume and temperature.

Pressure is defined as the force per unit area. It is calculated by dividing the force applied to an object by the area it acts on.

Key points

$$\text{pressure, } P = \frac{\text{force, } F}{\text{area, } A}$$

This means that pressure has units of newtons (force) per square metre (area): $N\,m^{-2}$.

This is also known as pascals (Pa).

Example

A block is laid on a bench as shown in Figure 2.2.

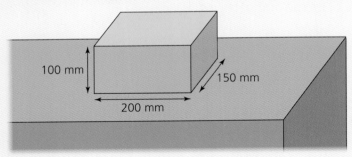

Figure 2.2

The block has a weight of 30 N. Calculate the pressure it exerts on the bench.

$F = 30\,N$

$A = 0.15 \times 0.2 = 0.03\,m^2$

$$P = \frac{F}{A} = \frac{30}{0.03}$$

$$= 1000\,N\,m^{-2}$$

$$= 1000\,Pa$$

Questions ?

The Relationships Sheet and Data Sheet will provide equations and values you will need. For example, $W = mg$.

1 **a)** A crate of 1500 kg with a base surface area of 0·9 m² is placed on the ground. Calculate the pressure this crate exerts on the ground.

   b) The crate is then placed on a smaller base, which has dimensions of 400 mm by 400 mm. What pressure does the base exert on the ground?

2 A person has a mass of 80 kg and a footprint of area of 250 cm². What pressure, in pascals, is exerted on the pavement by this person when walking?

3 A type of flooring can withstand a pressure of up to 12 000 Pa. A stool has a base of area 0·15 m². Calculate the maximum weight that the stool could support without damaging the floor?

4 A floor shows signs of damage where the legs of chairs seem to have dented or gouged a section. Explain, in terms of pressure, how this could have occurred.

5 Give a reason why footballers use studs or blades on the bottoms of their boots.

6 How do large transport trucks carry such heavy loads and yet not sink into the road?

The kinetic model

The kinetic model is a theory that explains the properties of gases – pressure, volume and temperature – in terms of the particles of the gas. The key concept is that the particles of a gas are continually moving (Figure 2.3). Their motion is random and rapid.

Daniel Bernoulli first proposed an accurate description of this theory in the 1730s but others, including the Scottish scientist Robert Brown, also put forward evidence in support of it.

Gas is the state of matter in which the particles are free to move and have only weak or no forces of attraction between them. They move freely in whatever space they occupy and collide randomly with each other. They also collide with the walls of their container. The kinetic model describes the properties of a gas in terms of its moving particles.

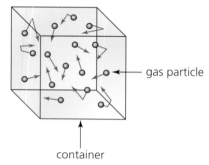

gas particle

container

Figure 2.3 The particles of a gas are continually moving

Pressure

The pressure of a gas is caused by the particles of the gas colliding with the walls of the container or any object in contact with the gas (Figure 2.4). If the particles make more collisions with the container walls, this results in an increase in pressure.

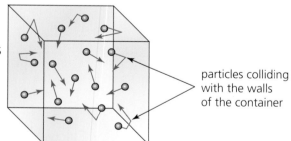

particles colliding with the walls of the container

Figure 2.4 Pressure in a gas is caused by particles colliding with the container walls

Temperature

The temperature of a gas is related to the average kinetic energy of the particles (Figure 2.5). The term average (or mean) is used because the motion of the molecules is random and some move more quickly than others. An increase in temperature means that the average velocity of the particles increases.

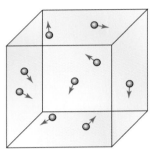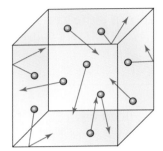

Gas at low temperature – particles are moving slowly

Gas at high temperature – particles are moving quickly

Figure 2.5 The particles of the gas at a higher temperature have more kinetic energy

Volume

The volume of a gas is the space it occupies. A certain amount of gas in a small container will 'expand' to fill a larger container if it is moved there (Figure 2.6). If a jar containing a gas is opened, the gas will move (diffuse) to spread out evenly around the space it now occupies.

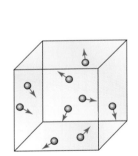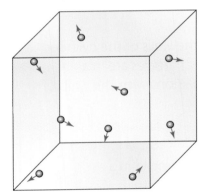

Particles close together in small container

Particles spread out in large container

Figure 2.6 The particles are close together in the small container but spread out in the larger container

The various properties of the gases are all connected. The kinetic theory provides a way to predict what will happen to these properties under certain circumstances.

The gas laws

Pressure and volume

An experiment was undertaken where a mass of gas in a cylinder was connected to a pressure gauge (Figure 2.7).

The piston in the syringe was moved to change the volume of the gas. The gauge measured the pressure as the gas was compressed. The temperature was kept constant. The results in Table 2.1 were obtained.

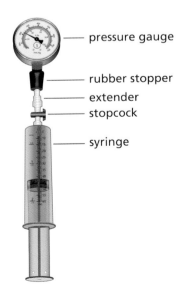

pressure gauge

rubber stopper

extender

stopcock

syringe

Figure 2.7 Investigating pressure and volume in a gas

Table 2.1 Results

Pressure (Pa)	101 000	121 000	152 000	202 000	303 000
Volume (cm³)	12	10	8	6	4

It was noted that the pressure increased as the volume of the gas decreased.

This leads to the relationship $P \times V =$ constant. This applies if the gas is held at a constant temperature.

Example

The gas in a cylinder is at a pressure of 120 000 Pa and occupies a volume of 18 cm³. The cylinder is compressed to a volume of 14 cm³. The temperature is kept constant. Calculate the new pressure.

$P_1 = 120\,000\,\text{Pa}$

$V_1 = 18\,\text{cm}^3$

$V_2 = 14\,\text{cm}^3$

$P_2 = ?$

$$P_1 \times V_1 = P_2 V_2$$

$$120\,000 \times 18 = P_2 \times 14$$

$$P_2 = \frac{120\,000 \times 18}{14}$$

$$= 154\,000\,\text{Pa}$$

$$= 150\,000\,\text{Pa}$$

Explanation using kinetic theory

Pressure is caused by the particles colliding with the walls of the cylinder. We compress the gas so that the same number of particles are in a smaller volume. This means there will be more collisions with the walls of the cylinder, increasing the pressure.

Pressure and temperature

A syringe of gas connected to a pressure gauge was placed in a temperature-controlled environment. The temperature was varied and the pressure noted. The volume was kept constant.

Typical results are shown in Figure 2.8. These results show that when the temperature was increased, the pressure of the gas also increased.

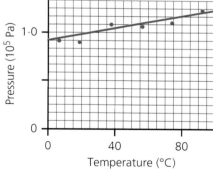

Figure 2.8 Investigating pressure and temperature in a gas

The graph did not pass through the origin, however. The graph was then extrapolated by continuing the line until it crossed the *x*-axis (Figure 2.9). This occurred at a temperature of −273 °C. This is known as **absolute zero**. It is the coldest possible temperature.

If we use this point of absolute zero as the start of our temperature scale, we have the **Kelvin scale**, measured in units of K. (See page 51 for more on the Kelvin scale and how to convert between Kelvin and Celsius.)

If we use the Kelvin scale we find that pressure and temperature increase in proportion. An increase in temperature causes a resultant increase in pressure when the volume is kept constant.

This leads to the relationship: $\dfrac{P}{T(\text{in } K)} = \text{constant}$

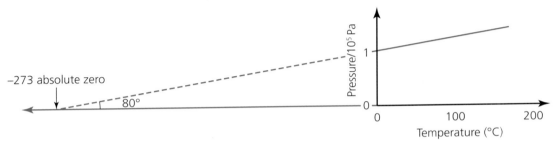

Figure 2.9

Example

A fixed mass of gas at 1.02×10^5 Pa is heated from 20 °C to 90 °C. The volume is constant. Calculate the new pressure.

$P = 1.02 \times 10^5$ Pa

$T_1 = 20\,°C = 293\,K$

$T_2 = 90\,°C = 363\,K$

$P_2 = ?$

$\dfrac{P}{T(\text{in } K)} = \text{constant}$

$\dfrac{1.02 \times 10^5}{293} = 348$

Use this as the 'constant' and now insert the new temperature:

$\dfrac{P_2}{363} = 348$

$\Rightarrow \quad P_2 = 348 \times 363 = 1.26 \times 10^5$ Pa

Explanation using kinetic theory

Pressure is caused by the particles colliding with the walls of the container. If we increase the temperature of the gas we increase the particles' kinetic energy and the particles move more quickly. This increase in motion causes more collisions with the walls of the container, increasing the pressure.

Volume and temperature

A syringe of gas, sealed at the nozzle but with the plunger free to move (see Figure 2.10), was placed in a temperature-controlled environment. The temperature was increased and it was noted that the volume of the gas also increased. The increase in temperature caused the gas to expand and the plunger to move out along the syringe, increasing the volume.

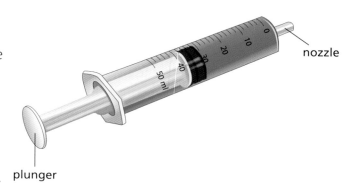

nozzle

If we use the Kelvin temperature scale we find that the volume and temperature increase in proportion. The increase in temperature causes a resultant increase in volume.

plunger

Figure 2.10 Investigating volume and temperature in a gas

This leads to the relationship: $\dfrac{V}{T\,(in\,K)} = $ constant

Example

The volume of a gas in a sample is $210\,cm^3$ at a temperature of $20\,°C$. The sample is put into a freezer at $0\,°C$. What volume will it occupy now?

$V_1 = 210\,cm^3$

$T_1 = 20\,°C = 293\,K$

$T_2 = 0\,°C = 273\,K$

$V_2 = ?$

$\dfrac{V}{T\,(in\,K)} = $ constant

$\dfrac{210}{293} = 0.7167$

Use this as the 'constant' and now insert the new temperature:

$\dfrac{V_2}{273} = 0.7167$

$\Rightarrow \quad V_2 = 273 \times 0.7167 = 196\,cm^3$

When the gas is heated the particles move more quickly, increasing the number of collisions with the walls of the container. As the plunger is free to move, the increased number of collisions pushes the plunger out. This increases the volume. The pressure exerted by the gas remains the same, because the volume is able to change instead.

Questions

1 A test tube containing $35\,cm^3$ of air at $21\,°C$ is placed in an oven and it expands to $48\,cm^3$. Calculate the temperature of the oven.

2 A sample of gas at $34\,°C$ is placed in a freezer and its volume reduces from $120\,cm^3$ to $92\,cm^3$. Calculate the temperature of the freezer.

Absolute zero and the Kelvin and Celsius temperature scales

Absolute zero is theoretically the lowest temperature that can be reached. It can be thought of as the temperature an object will reach when all of its thermal (kinetic) energy has been removed and we can remove no more.

Absolute zero is equivalent to −273 °C. This is 0 K on the Kelvin temperature scale.

To convert from Celsius to Kelvin we add 273.

To convert from Kelvin to Celsius we subtract 273.

The scales are similar in that an increase in temperature of 1 K is the same as an increase of 1 °C. The only real difference is the starting points.

The Kelvin scale is a better temperature scale than the Celsius scale because it starts from the lowest possible value and has no negative values (similar to mass, length and time). It does not give an easy indication of what the temperature will feel like, however, so the Celsius scale is still used most of the time.

Questions ?

1 Copy and complete the following table.

Celsius		−200		0		150		500
Kelvin	0		200		300		600	

2 A cyclist inflates a tyre to a pressure of 4.50×10^5 Pa at a temperature of 4 °C. After a long cycle the temperature inside the tyre has increased to 31 °C. Calculate the new pressure in the tyre.

3 A syringe of gas has a volume of 35 cm^3 at a pressure of 1.02×10^5 Pa. The volume of the syringe is increased to 72 cm^3. The temperature remains constant. Calculate the new pressure.

4 The results of an experiment to measure the volume and temperature of a fixed mass of gas are given below. Some results are missing. Copy and complete the table.

Volume (cm^3)	210	240	260			360	500
Temperature (°C)		0		50	80		

5 A diving tank contains 0.4 m^3 of air at 6.50×10^5 Pa. What volume would this air occupy at a normal air pressure of 1.01×10^5 Pa?

6 A driver notices that the pressure in the tyres of her car increases after the car has been used for a long drive. Using the kinetic theory, suggest a reason why the pressure increases.

Chapter 3: Waves

3.1
Properties of waves

What you should know

For N5 Physics you need to be able to:

★ understand that energy can be transferred as waves
★ determine the frequency, wavelength, amplitude and wave speed in appropriate situations
★ use appropriate relationships involving frequency, wavelength, distance and time
★ explain diffraction and its effect on long and short waves.

A wave is a term used in Physics to describe how energy is transferred from one point to another. Examples of waves are to be found in everyday life. Radio waves and microwaves are examples with which you are probably familiar.

You should understand terms related to waves, such as **wavelength**, **frequency**, **amplitude** and **period**.

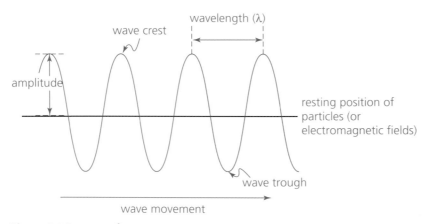

Figure 3.1 Features of a wave

There are two main types of waves, **transverse** and **longitudinal**. In a transverse wave the vibrations occur perpendicular to the direction of the wave. Figure 3.1 (page 52) shows a transverse wave. In a longitudinal wave the particles vibrate parallel to the direction in which the wave is travelling. A longitudinal wave is shown in Figure 3.2.

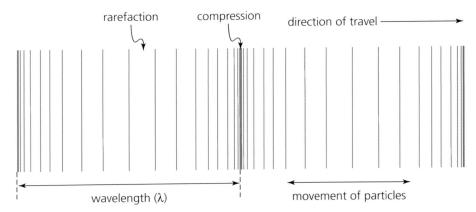

Figure 3.2 A longitudinal wave

In sound (longitudinal), seismic (longitudinal) and water (transverse) waves, the energy is transferred by the particles of air, earth or water vibrating. In most other waves, including those in the electromagnetic spectrum (transverse), electric and magnetic fields oscillate to transfer the energy. There is more about the electromagnetic spectrum in Topic 3.2.

The wave speed, frequency, wavelength and distance travelled are linked by simple relationships. Two important ones are:

$$\text{speed} = \text{frequency} \times \text{wavelength or } v = f \times \lambda$$

$$\text{speed} = \frac{distance}{time} \text{ or } s = \frac{d}{t}$$

The speed of all electromagnetic radiations, including light, is $3 \times 10^8 \, \text{m s}^{-1}$. You will need to use this in many wave calculations.

Examples

1 A sound wave of frequency 256 Hz has a wavelength of 1·3 m. What value does this give for the speed of sound?

$f = 256\,\text{Hz}$ $v = f \times \lambda$

$\lambda = 1·3\,\text{m}$ $\Rightarrow 256 \times 1·3 = 332·8\,\text{m s}^{-1} = 330\,\text{m s}^{-1}$

2 Red light has a wavelength of $7·04 \times 10^{-7}$ m. Calculate its frequency.

$v = 3 \times 10^{8}\,\text{m s}^{-1}$ $v = f \times \lambda$

$\lambda = 7·04 \times 10^{-7}\,\text{m}$ $\Rightarrow 3 \times 10^{8} = f \times 7·04 \times 10^{-7}$

$$f = \frac{3 \times 10^{8}}{7·04 \times 10^{-7}} = 4·26 \times 10^{14}\,\text{Hz}$$

Questions

1 A student has a range of hearing from 30 Hz to 17 500 Hz. Calculate the wavelengths of these sounds. The speed of sound is 340 m s^{-1}.

2 A dog can hear a sound at 25 000 Hz. Calculate the wavelength of this sound. The speed of sound is 340 m s^{-1}.

3 An ultrasound transmitter transmits waves of frequency 5 MHz. In soft human tissue the speed of sound is 1500 m s^{-1}. Calculate the wavelength of these ultrasound waves.

4 Microwaves are electromagnetic waves. They have a wavelength of 0·06 m. Calculate their frequency.

5 Mobile phones use microwaves. They can operate on frequencies of 1800 MHz and 1900 MHz. Calculate the associated wavelengths of these frequencies.

Waves have a number of properties that identify them. These are **reflection**, **refraction**, **interference** and **diffraction**.

Reflection occurs when a wave strikes a barrier and is reflected off that barrier. A familiar example is when a sound produces an echo.

Refraction occurs when a wave changes speed as it moves from one medium to another. There is more about refraction in Topic 3.3.

Interference occurs when two or more waves combine.

Diffraction is a property of a wave. When waves pass by a barrier or by an edge or through a gap they spread or bend behind the barrier. Examples of diffraction are shown in Figure 3.3.

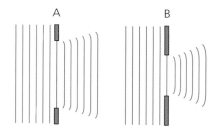

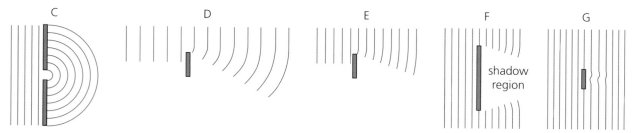

Figure 3.3 Examples of waves diffracting around barriers and through gaps

When a gap is approximately the same size or less than the wavelength semi-circular waves are formed on the other side of the gap. The frequency, wavelength and speed of the waves that have diffracted do not change.

From parts D and E of Figure 3.3, a difference in the pattern of diffraction can be noted. Waves with larger wavelengths diffract more than waves with smaller wavelengths.

A simple example of this is found at a music concert. If you move behind a pillar you can still hear the music while not seeing the performer. Sound waves have a larger wavelength than light waves and so diffract more.

Another example of this is music coming from an event around the corner from you. You would hear the lower bass sounds more than the higher sounds. The low notes have a larger wavelength and diffract more, so they can be heard further around the corner.

Diffraction also occurs in radio waves. Longer wavelengths diffract around objects like hills or large buildings but shorter wavelengths cannot (see Figure 3.4). Reception of these shorter wavelength signals may therefore not be very good if the receiver is in the shadow of a large structure. Reception in hilly or remote areas is often more difficult for shorter wavelength signals, as they do not diffract over and around the landscape.

A similar effect is occasionally found in cities with high buildings that block the signals from GPS satellites, for example. These areas are sometimes described as 'urban canyons'.

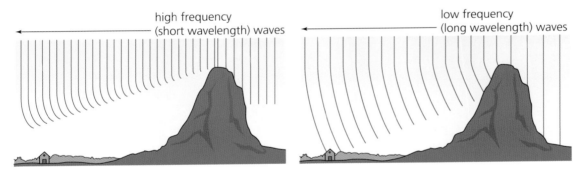

Figure 3.4 Longer wavelengths diffract around obstacles but shorter ones cannot. The shorter wavelength is blocked by the hill; only the longer wavelength reaches the house

3.2
Electromagnetic spectrum

What you should know

For N5 Physics you need to be able to:
* ★ describe the relative frequencies and wavelengths of bands of the EM spectrum
* ★ describe some typical sources and applications of bands of the EM spectrum
* ★ understand that the energy associated with each band is related to the frequency
* ★ state that all radiations in the EM spectrum travel at the speed of light.

The electromagnetic (EM) spectrum includes all the various forms of electromagnetic radiation that exist (see Figure 3.5). The term **spectrum** refers to the way in which we order them: by increasing wavelength or increasing frequency, for example.

All the different types of wave in the EM spectrum are examples of the same type of radiation (electromagnetic), so they all travel at the same speed. This is the speed of light: $3 \times 10^8 \, \mathrm{m \, s^{-1}}$.

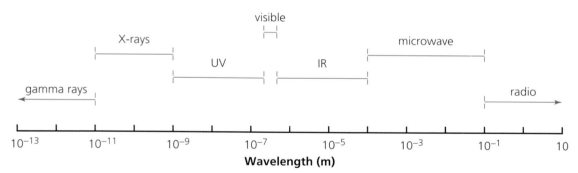

Figure 3.5 The electromagnetic spectrum

Roughly in the middle of this electromagnetic spectrum lies the visible spectrum (the range of light that can be seen, Figure 3.6).

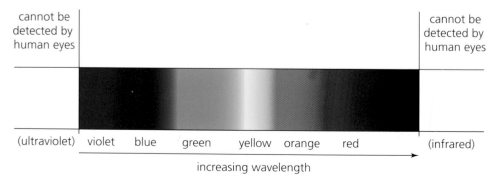

Figure 3.6 The visible spectrum

Parts of the electromagnetic spectrum

Radio waves

Radio waves have the longest wavelength, from around 0·1 m to several kilometres. Due to their size they have the most detectable wave characteristics (diffraction, for example). They are used in telecommunications throughout the world and in astronomy.

Microwaves

Microwaves have wavelengths between about 10^{-4} m and 0·1 m. They are used in telecommunications, Wi-Fi, Bluetooth, medicine, astronomy and home appliances.

Infrared

Infrared light has a wavelength between about 10^{-4} m and 10^{-7} m. It is used in astronomy, spectroscopy, night vision, telecommunications and military devices. It is referred to as heat or thermal energy in some cases.

Visible light

Visible light has a wavelength of $400-700 \times 10^{-7}$ m. This is the range of colours we can see.

Ultraviolet

Ultraviolet light has a wavelength between about 10^{-7} m and 10^{-9} m. It is found in energy from the Sun and can cause skin discoloration and burning. UV light is widely used in a range of scientific processes and to detect counterfeit bank notes.

X-rays

X-rays have wavelengths from about 10^{-9} m to 10^{-11} m. They are widely used in medical situations, both for diagnosis and for treatment of certain conditions. They are also emitted by various stellar objects.

Gamma rays

Gamma rays have wavelengths from about 10^{-11} m to 10^{-13} m. This is the smallest wavelength of EM radiation. They are emitted by certain radioactive isotopes and by highly energetic stellar objects such as supernovae, pulsars and quasars.

All electromagnetic radiation travels at the speed of light: 3×10^8 m s^{-1}. This allows us to calculate the frequency of the radiation if we know the wavelength.

1 A band of infrared radiation has a wavelength of $2{\cdot}0 \times 10^{-5}$ m. Calculate its frequency.

$\lambda = 2{\cdot}0 \times 10^{-5}$ m $\qquad\qquad v = f \times \lambda$

$v = 3 \times 10^{8}$ m s^{-1} $\qquad \Rightarrow \quad 3 \times 10^{8} = f \times 2{\cdot}0 \times 10^{-5}$

$f = ?$ $\qquad\qquad \Rightarrow f = \dfrac{3 \times 10^{8}}{2 \times 10^{-5}} = 1{\cdot}5 \times 10^{13}$ Hz

This band of infrared radiation has a frequency of $1{\cdot}5 \times 10^{13}$ Hz.

2 Gamma rays have a wavelength of $1{\cdot}5 \times 10^{-11}$ m. Calculate their frequency.

$\lambda = 1{\cdot}5 \times 10^{-11}$ m $\qquad\qquad v = f \times \lambda$

$v = 3 \times 10^{8}$ m s^{-1} $\qquad \Rightarrow \quad 3 \times 10^{8} = f \times 1{\cdot}5 \times 10^{-11}$ m

$f = ?$ $\qquad\qquad \Rightarrow f = \dfrac{3 \times 10^{8}}{1{\cdot}5 \times 10^{-11}} = 2{\cdot}0 \times 10^{19}$ Hz

Energy and frequency

The frequency of the electromagnetic spectrum varies from about 3×10^{7} Hz to 3×10^{20} Hz. The energy associated with the different form s of EM radiation is in direct proportion to the frequency. Radio waves have much less energy than X-rays, for example.

The increased energy of higher frequencies also increases the associated health risks. For example, there have been claims that excessive use of mobile phones (which use microwaves) can cause cell or tissue damage, but no studies have been able to provide proof of this. UV, X-rays and gamma rays, however, are definitely harmful and over-exposure to these types of radiation can have severe health implications.

1 Use Figure 3.5 to select one wavelength value for each of the seven forms of EM radiation. Calculate the frequency associated with each of your chosen wavelengths.
2 Research and write a short paragraph on each of the following:
   a) the use of gamma rays in medicine
   b) applications of thermal imaging
   c) the risks and benefits of ultraviolet radiation.
3 List the range of frequencies of X-rays and describe two of their possible uses.
4 Should tanning salons be allowed to open freely? Justify your opinion.
5 If X-rays are harmful, why do we allow them to be used?
6 Have mobile phones been shown to be harmful?

3.3
Light

What you should know

For N5 Physics you need to be able to:

★ explain refraction in terms of change of wave speed
★ identify appropriate angles and the normal in refraction diagrams.

Light is a wave and it exhibits wave characteristics. **Refraction** occurs when light passes from one material (or medium) into another – from air to glass, for example. When light travels from air into glass, its speed changes.

The speed of light in a vacuum is $3 \times 10^8 \, \mathrm{m \, s^{-1}}$. In air its speed is *almost* the same, $3 \times 10^8 \, \mathrm{m \, s^{-1}}$; it is actually very slightly slower but we usually take it to be the same.

Table 3.1 Speed of light in different materials

Medium	Speed of light ($\mathrm{m \, s^{-1}}$)
air	$3{\cdot}00 \times 10^8$
glass	$1{\cdot}87 \times 10^8$
diamond	$1{\cdot}24 \times 10^8$
water	$2{\cdot}25 \times 10^8$

This change in the speed of light, when it passes from air to glass, for example, causes the light to change direction when it strikes the glass at an angle (Figure 3.7).

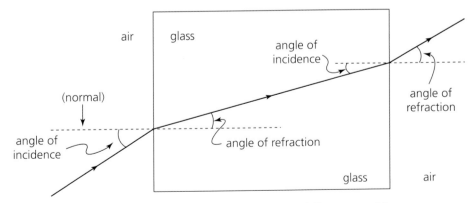

Figure 3.7 Refraction of a ray of light as it moves between different materials

* The **normal** is an imaginary line drawn perpendicular to the boundary between the two materials.
* The **angle of incidence** is the angle between the incident ray (before it is refracted) and the normal.
* The **angle of refraction** is the angle between the refracted ray and the normal.

When moving from air (less dense) to glass (more dense), the light changes direction slightly **towards** the normal. This can be described as:

the angle of incidence > the angle of refraction

When going from glass (more dense) to air (less dense), the light changes direction **away** from the normal. This can be described as:

the angle of incidence < the angle of refraction

Key point ❗

* The density of a material when used in the context of refraction refers to the material's 'optical density'. This is not the same as our usual meaning of density, which is mass per unit volume.

Most applications of refraction, such as cameras, projectors, spectacles and so on, rely on the fact that light changes **direction** slightly when moving from one medium to another. However, the definition of refraction is that the light changes **speed** when moving from one medium to another.

Lenses

Lenses are pieces of glass or Perspex that are shaped in a certain way in order to alter the direction of light to suit our purposes. There are two main types of lens, converging and diverging, as shown in Figure 3.8.

(a)

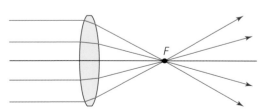

(b)

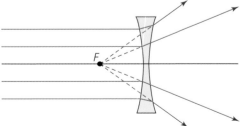

Figure 3.8 (a) Converging lens; **(b)** Diverging lens

The shape of the lens changes the direction of the rays of light at the first boundary (air to glass) and also at the second boundary (glass to air). Shaping the glass allows us to manipulate the direction of the light as we wish.

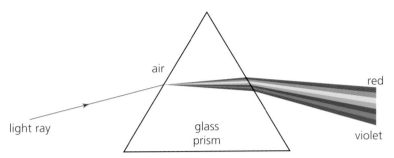

Figure 3.9 Prism being used to split light into its spectrum of colours

In a glass prism the light refracts at the first boundary. The colours that make up white light have different **refractive indices** and so they refract by slightly different amounts. This splits the white light very slightly at the first boundary. The 'split' light travels to the second boundary, where refraction occurs again, exaggerating the split. The light leaving the prism can now be seen as a spectrum of distinct colours (Figure 3.9).

This happens in raindrops to create rainbows.

Key point

* The refractive index is a measure of the change in velocity, and also of direction, that radiations experience when travelling from one medium to another.

Questions ?

1 Draw a diagram of a ray of light striking a glass block at an angle of 35 degrees to the normal. Show the path the light takes and indicate the normal, angle of incidence and angle of refraction.
2 Why do objects immersed in a glass of water appear to bend?
3 List three devices that rely on the refraction of light in order to operate.

Chapter 4: Radiation

Nuclear radiation

What you should know

For N5 Physics you need to be able to:

★ describe the nature of alpha, beta and gamma radiation, including the relative effects of ionisation, absorption and shielding
★ discuss background radiation sources
★ quantify the absorbed dose and equivalent dose and compare equivalent doses due to a variety of natural and artificial sources
★ describe applications of nuclear radiation
★ state that activity is measured in becquerels and use $A = \dfrac{N}{t}$ in problems
★ determine the half-life of a material and use graphical or numerical data in problems
★ give a qualitative description of fission and fusion, emphasising the importance of these processes in the generation of energy.

Radioactive materials and their effects were first discovered in 1896 by Henri Becquerel, who found that certain rocks could darken photographic plates. This work was continued by other notable scientists such as Curie and Rutherford. The effect was first discovered in rocks in which the 'active ingredient' was mainly uranium.

Radioactive materials found in the Earth were first created before the formation of the Solar System. When our planet came together some of these materials were embedded in the Earth. Many radioactive materials used now were formed from these original radioactive atoms.

Nuclear radiation is also known as **ionising radiation**. It is able to remove electrons from atoms to give charged **ions**. It is this ability to strip electrons from atoms that leads to the harmful properties of nuclear radiation.

There are three main types of nuclear radiation and they are all generated from changes in the **atomic structure** of an atom (see Figure 4.1). An atom has a nucleus that is composed of neutrons and protons. The electrons orbit the nucleus.

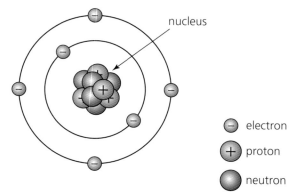

Figure 4.1 The atomic structure of an atom

Types of radiation

Alpha radiation, α

An alpha particle is a large particle composed of two protons and two neutrons (no electrons). This gives it an overall charge of +2. It is ejected from a radioactive nucleus at a fairly high speed. As it passes through materials it ionises their atoms and gradually loses its energy.

Due to its size and charge the alpha particle's energy is easily absorbed by the materials it meets, so it cannot penetrate deeply into them. As a result it cannot penetrate our skin and can cause us little harm externally. If ingested, however, all the energy is absorbed by the tissue surrounding it. This can be very harmful to us.

Beta radiation, β

A beta particle is an electron, so it has a charge of −1. It is ejected from a radioactive nucleus at high speed. Its smaller mass and charge mean it can pass further through a material before it loses its energy. Beta is less harmful than alpha if ingested, but its penetrating ability means it can be more harmful externally.

Gamma radiation, γ

Gamma radiation is a high-energy photon emitted from an excited or energetic atom. It has no charge and no mass. It can penetrate large distances through materials and is difficult to stop completely. It can cause ionisation by transferring some of its energy to atoms, releasing an electron.

Shielding

When working with radioactive materials, a worker needs to be protected from the rays and particles coming from the radioactive source. This is done by keeping the worker as far away from the source as possible, by limiting the time that the source is used, and by placing **shields** between the worker and the source.

The three types of radiation can be categorised by their penetrating ability (see Figure 4.2).

Alpha radiation is not very penetrating. It can be stopped by a sheet of paper or even a distance of 20 cm or so of air.

Beta radiation is more penetrating than alpha. It can pass through air and paper without being much reduced. It requires a thin sheet of a material such as aluminium to noticeably stop beta radiation.

Gamma radiation is very penetrating and passes through paper and aluminium with little or no reduction. Thick sheets of materials such as lead or concrete are required to reduce gamma radiation.

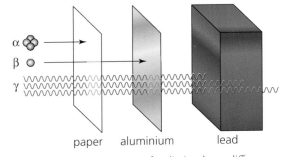

Figure 4.2 The three types of radiation have different penetrating abilities

Background radiation

Radioactive materials are found naturally. They occur in rocks, soil, water and so on. Uranium, for example, is an ore that is mined and refined to produce material for use in nuclear power stations. Radon gas is the largest contributor to background radiation. It is a product of uranium and the gas seeps through the soil and can reach high levels in basements with little ventilation.

There is also radiation caused by energetic particles coming from the Sun and other stars. This is known as **cosmic radiation**. These particles can arrive directly or can interact with atoms in the upper atmosphere to create a sort of 'secondary shower' of radiation. This means that the background radiation level in the upper atmosphere is greater than that on the ground.

Some rocks, such as granite, have higher levels of radiation than others. The soil or bedrock in an area with a lot of these rocks can increase the local background level.

Medical uses of radiation and nuclear power stations (see page 66) also contribute to the background radiation level.

Radiation doses

Radiation is definitely harmful to humans in large doses, but it is difficult to predict exactly what the effect of a small dose will be. It may harm some people but not others. There is often a great deal of misunderstanding and bad science regarding this question.

To give an indication of the effect of radiation on a body we use certain terms and relationships.

Absorbed dose

This is the energy absorbed from a radioactive material divided by the mass of the material receiving the dose:

$$D = \frac{E}{m}$$

Where D = absorbed dose, measured in grays (Gy) – 1 Gy = 1 Jkg^{-1}

$\qquad E$ = energy, measured in joules (J)

$\qquad m$ = mass, measured kg

This can be thought about in the same way as a drug or medicine, for example. The effect of a drug will be much greater if the same amount is given to an infant compared to a fully grown adult. The same applies with radiation. This relationship makes no reference to the **type** of radiation.

Equivalent dose

The effect of the type of radiation is introduced into the equation with the use of the **radiation weighting factor**, W_R.

Equivalent dose = absorbed dose × radiation weighting factor

$$H = D \times W_R$$

H is the **equivalent dose**, measured in sieverts (Sv).

The radiation weighting factor is a number that indicates the biological or harmful effect of the radiation.

Sample W_R values: $\alpha = 20$; $\beta = 1$; $\gamma = 1$

These values indicate that **for the same amount of energy received**, alpha radiation is 20 times more harmful than beta or gamma.

Table 4.1 Sample doses of background radiation

2·4 mSv/yr	Typical background radiation experienced by everyone (average 1·5 mSv in Australia, 3 mSv in North America)
Up to 5 mSv/yr	Typical incremental dose for aircrew in middle latitudes
10 mSv	Effective dose from abdomen and pelvis CT scan
50 mSv	Allowable short-term dose for emergency workers
250 mSv	Allowable short-term dose for workers controlling the 2011 Fukushima accident

The amount of exposure to radiation depends on a number of factors and in Britain the National Radiological Protection Board (NRPB) monitors and advises in these matters.

The average dose equivalent received annually due to background radiation in the UK is 2·2 mSv. As part of everyday life, however, we may receive small additional doses, from X-rays for example.

The effective dose limit for a member of the public is 1 mSv.

Workers in certain industries will receive larger doses and these are described in terms of effective dose. This is the equivalent dose received by the worker but adapted to describe an overall dose to the body.

The effective dose limit for a radiation worker is 20 mSv.

Questions

1 An 80 kg man and a 0·5 kg hamster each receive 0·02 J of energy from exposure to a radioactive source.
   a) Calculate the absorbed dose for each.
   b) It was found that the source emitted alpha radiation. Calculate the dose equivalent that they each received.
   c) Why has the hamster received a larger dose equivalent?
2 Draw an alpha particle and give a brief description of its composition, charge, penetration and possible harmful effects.
3 Explain why airline crews receive higher doses of radiation than people at ground level.
4 List some possible sources of background radiation.
5 Suggest why background radiation in Aberdeen is slightly higher than that in Glasgow.

Applications of nuclear radiation

There are many situations where nuclear radiation can be useful. They can be broadly categorised into the following:

- medical uses
- academic/scientific uses
- industrial uses
- energy generation
- military uses.

Medical uses

Medical uses range from imaging in X-rays and CAT scans to the treatment of certain conditions. For example, large doses of radiation are often given to kill cancerous cells and tumours. Tracers involve a radioactive material being injected or ingested (swallowed); the progress of the material can then be followed to search for a blockage, for example. Many hospitals have departments of Nuclear Medicine where doctors and scientists supervise the use of nuclear radiation for patient treatments.

Academic/scientific uses

Radioactive materials can be used to perform experiments or demonstrate principles in order to educate students. Scientists can also experiment with various types of radiation and doses to determine their effects on different materials. The age of an old piece of wood or cloth can be estimated by measuring the amount of a radioactive isotope called carbon-14 that it contains. This is known as carbon dating.

Industrial uses

There are many uses of nuclear radiation in industry. For example, smoke detectors contain a small amount of radioactive material. If smoke surrounds the alarm, the amount of ionisation due to the radioactive material is reduced and this is detected. This sets off the alarm.

Some firms use nuclear radiation to monitor the thickness of the materials they produce (from paper to plastic to metals). Fabric for clothing is irradiated prior to chemicals being added so that the chemicals (anti-wrinkle, for example) bind more securely and to remove bacteria from the fabric. Non-stick pans are irradiated with gamma radiation to make them function more effectively. In agriculture, many seeds are irradiated to ensure stronger and more disease-resistant crops.

Energy generation

Nuclear power is one of the main sources of energy used in developed countries. Uranium ore is mined, refined and used to supply large amounts of heat energy. This heat energy boils water, which becomes steam and turns turbines to generate electrical energy. Nuclear power contributes a significant amount to Britain's energy supplies. Nuclear reactors are also used to power nuclear submarines and some of our satellites. There is more about nuclear power later in this topic.

Military uses

Nuclear weapons are a fact of modern life but a concern for many. The energy released from an uncontrolled nuclear reaction can be immense and cause catastrophic devastation. While nuclear weapons have only ever been used in the Second World War, the political impact of a country having nuclear weapons capability is huge.

Example

Select one industrial use of nuclear radiation and describe its general principles.

Welding checking: Aircraft, for example, need to have their welds and joints checked regularly to ensure they are safe to fly. A gamma source is placed on one side of a weld or joint and a detector on the other side. The amount of radiation that passes through is checked against recorded data. If there is an increase in the level, this could be an indication of a crack or gap.

Activity of radioactive materials

Radioactive materials can be described by their **activity**. This is a general term that refers to the number of nuclear disintegrations in a given time.

When an atom emits an alpha particle, for example, the atom splits or disintegrates. The activity of a source is described in terms of the number of disintegrations per second.

If an isotope of uranium undergoes 24 000 disintegrations in a minute, its activity is determined by

$$A = \frac{N}{t} = \frac{24\,000}{60} = 400 \text{ becquerels (Bq)}$$

1 Bq = 1 disintegration per second

The activity of a radioactive material refers to the total number of disintegrations per second. In the laboratory this is difficult to measure and in schools we measure the activity of a source using a Geiger–Müller tube and a counter (Figure 4.3).

This doesn't really measure the activity as it can only detect the radiation that is directed at the tube. It is, however, indicative of the activity and is used in some cases to determine half-life for example.

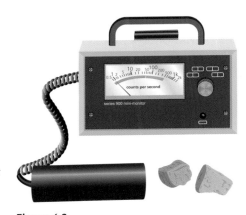

Figure 4.3

Activity and half-life

The activity of a radioactive substance can also be used to describe how its radioactivity varies with time.

The first scientists to study radioactivity found that the activity of a radioactive material reduced over time, but that this was not a simple process. Radioactive materials did not lose a certain amount of activity each year, for example.

It was found that graphs of activity against time for all radioactive materials had the same shape, similar to that in Figure 4.4. Further study showed that the time taken for the activity to reduce by half was always the same for a particular radioactive material. This time is known as the **half-life** of the material.

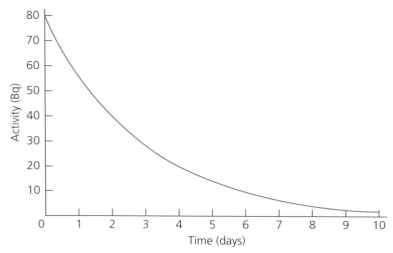

Figure 4.4 How the activity of a radioactive material decreases over time

To calculate the half-life of a material, we need to measure and plot its activity over a period of time. This gives a graph similar to the one in Figure 4.4.

A value for the activity is then chosen and the corresponding time is read off the graph. This initial activity value is then divided by two and the corresponding time is again read off the graph. The difference between the two times is the half-life for that material.

Example 🚩

Use the graph in Figure 4.4 to determine the half-life of a material.

First select a suitable value for the activity. On this graph, 40 Bq corresponds to a simple time that will be easy to work with, but any activity value can be chosen (see below).

Read off the time from the graph: 2 days in this case.

Calculate half of the activity chosen: 20 Bq.

Again read off the time: 4 days.

Time taken for activity to halve is 4 days − 2 days = 2 days.

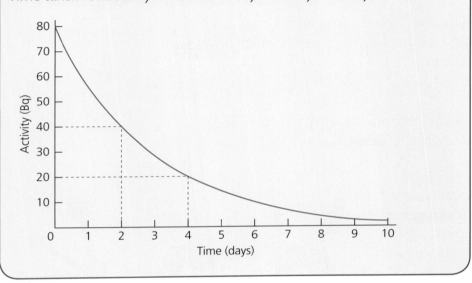

Analysing half-life

The half-lives of different radioactive materials have a wide range of values, as shown in Table 4.2.

Table 4.2 Examples of half-lives

Radioisotope	Half-life
Polonium-215	0·0018 seconds
Bismuth-212	60·5 seconds
Sodium-24	15 hours
Iodine-131	8·07 days
Cobalt-60	5·26 years
Radium-226	1600 years
Uranium-238	4·5 billion years

Some materials decay quickly while others take millions of years. For example, iodine-131, which is used as a tracer, has a half-life of 8·07 days. When this is injected into a patient, we know that after 8 days there will be half of the radioactive iodine remaining, after 16 days a quarter, after 24 days one-eighth, and so on. Materials with a short half-life soon decay to a level that is less significant than background radiation.

Radium-226, on the other hand, has a half-life of 1600 years and will remain radioactive for a very long time.

The storage and disposal of radioactive waste is a complex and difficult issue. For materials with long half-lives, the radiation will be significant for a long time. Current methods rely on processing the waste into very hard and stable discs and then burying these discs deep underground in rocks that appear to withstand radiation. People living near to these sites are generally unhappy about these methods, however.

Nuclear energy

Nuclear power stations use radioactive materials to generate large amounts of heat. This heat is used to turn water into steam, which drives turbines to generate electrical energy.

Nuclear reactions

There are two main types of nuclear reaction – fission and fusion. The reaction used in nuclear power stations is **fission**.

Fission reaction

Fission means 'to split'. In a nuclear reactor, uranium fuel rods are placed close together. We then cause the uranium atoms to split (Figure 4.5) by bombarding the nucleus with neutrons. When the atoms split, large amounts of energy are released and used to heat water.

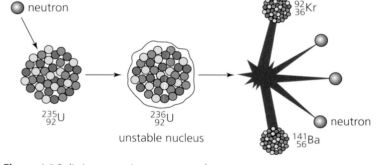

We can show the splitting of the uranium atom using an equation like this:

$$^{235}_{92}U + {}^{1}_{0}n \rightarrow {}^{141}_{56}Ba + {}^{92}_{36}Kr + 3{}^{1}_{0}n + energy$$

Figure 4.5 Splitting a uranium atom to release energy

The main reaction is one where uranium is 'hit' with a neutron. This causes the uranium to split, creating two smaller nuclei and three neutrons. These neutrons can then go on to split other uranium atoms, again producing more neutrons. This is an example of a **chain reaction**.

The energy is generated due to a reduction in the mass of the materials. There is slightly more mass before the reaction than after the reaction. This difference in mass is converted to energy. The amount of energy can be calculated using the relationship:

$$E = mc^2$$

Where E = energy released

m = mass lost

c = speed of light ($3 \times 10^8 \, m \, s^{-1}$)

The energy produced by splitting one atom is very small – about 3×10^{-11} J. There are vast numbers of atoms that can split, however. This allows power of the order of hundreds of megawatts or even gigawatts to be produced.

Fusion reaction

A **fusion** reaction is in some ways the opposite of a fission reaction. It is when two smaller nuclei combine to form a larger nucleus.

Deuterium and tritium atoms (isotopes of hydrogen) can be 'fused' together under special conditions to form a helium atom and a neutron (Figure 4.6). Energy is also released due to an overall decrease in the mass after the reaction.

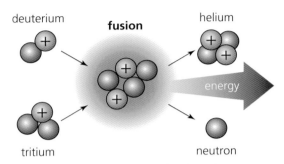

Figure 4.6 A fusion reaction using deuterium and tritium

We can show the fusion of deuterium and tritium using an equation like this:

$$^3_1H + {}^2_1H \rightarrow {}^4_2He + {}^1_0n$$

This is an example of the kind of reaction that occurs in stars like our Sun. Smaller nuclei (mainly hydrogen) fuse together to form larger nuclei (mainly helium).

The problem with humans using nuclear fusion to generate energy is the conditions needed for fusion to occur. The atoms must be heated to temperatures of about 40 000 000 Kelvin or be pressurised enough so that they become close enough together for fusion to occur. (We can induce fusion reactions but this uses more energy to start than it gives out.)

When atoms are heated to such temperatures they become a different state of matter, called **plasma**. Containing this plasma is difficult. It can be done using very strong magnetic fields but it is technically complex and expensive.

ITER is an international project where an experimental prototype fusion reactor is being designed and constructed in France. It involves building a tokamak (a large magnetic ring) to contain the plasma. It is hoped that it will be able to produce 500 MW of power for every 50 MW of power used to initiate and sustain the temperatures for the reaction.

The possible benefits are huge. The fuel required is abundant and the plant would have very low CO_2 emissions. There is only very minor radioactive contamination and no chance of a nuclear reaction disaster similar to Chernobyl or Three Mile Island.

Questions

1. List and describe two industrial uses of nuclear radiation.
2. List and describe two medical uses of nuclear radiation.
3. A radioactive isotope has a mass of 64 g and a half-life of 30 days.
   a) How much will have decayed after 90 days?
   b) How much of the original isotope will remain after 120 days?
4. A radioactive source has an activity of 40 000 Bq and a half-life of 8 hours. Draw a graph of its activity against time over a period of one day.
5. What is the main fuel used in Britain's nuclear power stations?
6. What would be the main fuel in a fusion power station?
7. Twenty years ago nuclear power stations were seen as dirty and polluting, and oil power stations as a cleaner source of energy. This has now changed significantly. Explain why nuclear power is seen as cleaner in some respects than coal- and oil-fuelled power stations.
8. What are the problems associated with the storage of nuclear waste?
9. The activity of a source was measured over a period of time. The results were plotted on a graph as shown in Figure 4.7.

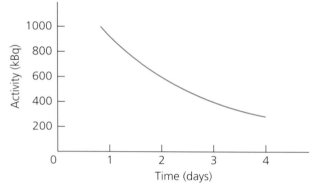

Figure 4.7

   a) Calculate the half-life of the source.
   b) Calculate the activity of the source at day 6.
10. A radioactive tracer has a half-life of 30 minutes. It has an initial activity of 9·6 kBq. Calculate the time taken for its activity to decrease to 600 Bq.
11. Carbon-14 has a half-life of 5730 years. Its initial activity is 36 mBq. How long will it take for its activity to fall to 2·25 mBq?

Chapter 5: Dynamics

Describing movement

What you should know

For N5 Physics you need to be able to:

★ understand and use the terms distance, displacement, time, speed, average speed, velocity, average velocity and acceleration
★ classify physical quantities as scalars or vectors
★ understand and use the relationships $d = vt$, $d = \bar{v}t$, $a = \frac{\Delta v}{t}$ and $a = \frac{v - u}{t}$
★ describe experiments to measure instantaneous and average speed
★ describe an experiment to measure acceleration.

Key points

∗ A **scalar** quantity has size only.
∗ A **vector** quantity has both size and direction.

Distance and displacement

When an object moves, the distance travelled is measured along the path followed by the object. Distance is a **scalar**. The SI unit of distance is the **metre** (m) and the symbol for distance used by SQA is d (or s).

By contrast, the displacement of the object is measured **in a straight line** from its start point to its finish point. Displacement is a **vector** – it has both size and direction. The SI unit for displacement is also the **metre** (m) and the symbol for displacement used by SQA is also d (or s).

When an object travels in a straight line **without changing direction**, the distance it travels and its displacement are numerically equal.

Example

An object starts at A. It moves in straight lines first to B, then to C, then to D and finally to E, where it stops (Figure 5.1).

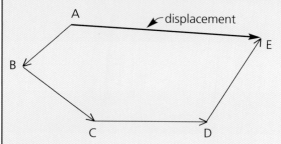

Figure 5.1

a) How is the distance moved by the object calculated?
The distance moved by the object $d = AB + BC + CD + DE$.

b) How is the displacement of the object calculated?
The displacement is represented by the line AE.

The size of the displacement is the length of AE, i.e. $d = AE$.
The direction of the displacement is the direction of line AE.

Time, speed and velocity

Time is a scalar. The SI unit for time is the **second** (s) and the symbol for time used by SQA is t.

Speed is a measure of how quickly an object travels a certain distance. It is a scalar. The SI unit of speed is **metre per second** ($m\,s^{-1}$) and the symbol for speed used by SQA is v.

The relationship between distance, speed and time is
distance = speed × time.

Velocity measures the speed **and** direction of an object. Velocity is a vector. The SI unit of velocity is also **metre per second** ($m\,s^{-1}$) and the symbol for velocity used by SQA is also v.

The relationship between displacement and velocity is
displacement = velocity × time.

When an object travels in a straight line **without changing direction**, its speed and velocity are numerically equal.

Average speed and average velocity

The average speed of an object is calculated by dividing the **total distance** it travels by the **total time** taken. The time may include periods when the object is stopped, moving slowly, moving quickly or changing speed. The SI unit of average speed is **metre per second** ($m\,s^{-1}$) and the symbol for average speed used by SQA is $\bar{v}$ (vee bar).

Average velocity is calculated by dividing the **final displacement** by the **total time** taken. The time may include periods when the object is stopped, moving slowly, moving quickly or changing velocity. The SI unit of average velocity is also **metre per second** ($m\,s^{-1}$) and the symbol for average velocity used by SQA is also $\bar{v}$.

Example

A girl walks 0·96 km from home to school in a time of 12 minutes. Calculate her average speed on her way to school.

$$d = 0.96\,km = 960\,m$$
$$t = 12\,min = 720\,s$$

$$d = \bar{v}t$$
$$\Rightarrow \quad 960 = \bar{v} \times 720$$
$$\Rightarrow \quad \bar{v} = 1.3\,m\,s^{-1}$$

Experiment: Measuring speed

To measure the speed of an object, you have to measure the distance, d, which it has travelled and the time, t, it has taken. You can choose the equipment you use to measure these quantities. For example, you can measure distance using a ruler, metre stick or tape measure, depending on the distance to be measured. Similarly, you can use a stopwatch, stopclock or a light gate and a computer, depending on the time to be measured. Other measuring devices may also be used.

It is good experimental practice to repeat each measurement **five** times and then calculate an average value. This improves the accuracy of the measurements.

To find the speed, you then use the relationship:

$$v = \frac{d}{t}$$

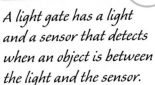

Hints & tips

A light gate has a light and a sensor that detects when an object is between the light and the sensor.

Example

Two groups of students were asked to measure the speed of a remote-controlled toy car. Group 1 measured the **instantaneous speed** of the car as it passed a point on the floor. Group 2 measured the **average speed** for the car to travel the length of the room. Every student took one of the measurements.

Group 1 used a ruler to measure the length of the car and a computer to measure the time for the car to pass through a light gate on the floor. Group 2 used a metre stick to measure the length of the room and a stopwatch to measure the time for the car to travel this distance.

Here are the results obtained by the students. $\Rightarrow$

Group 1		Group 2	
Distance (cm)	Time (s)	Distance (m)	Time (s)
15·1	0·51	12·02	46·6
14·9	0·49	11·99	45·7
15·0	0·50	12·04	48·4
15·2	0·49	12·03	47·4
14·8	0·51	11·97	45·3

Average distance = 15 cm = 0·15 m
Average time = 0·50 s
$$v = \frac{d}{t} = \frac{0·15}{0·5} = 0·30 \, \text{m s}^{-1}$$
This is an instantaneous speed as it is measured over a short time and distance.

Average distance = 12·01 m
Average time = 46·68 s
$$v = \frac{d}{t} = \frac{12·01}{46·68} = 0·257 \, \text{m s}^{-1}$$
This is an average speed as it is measured over a longer time and distance.

Suggest a possible reason why the measured instantaneous speed of the toy car is greater than the measured average speed of the toy car.

Group 1 may have measured the instantaneous speed when the car had reached its maximum speed while Group 2 may have measured the average speed of the car from a standing start.

Acceleration

Acceleration measures how quickly the velocity of an object is changing. Acceleration is a vector. The SI unit of acceleration is **metre per second squared** (m s^{-2}) and the symbol for acceleration used by SQA is a.

Acceleration is defined as change in velocity in unit time, $a = \frac{\Delta v}{t}$ (the symbol Δ, delta, means 'change in').

When an object has **constant acceleration** from **initial velocity** u to **final velocity** v in a time t, the acceleration is given by $a = \frac{v - u}{t}$.

Hints & tips

When you use this equation you **must** make sure you substitute the initial and final velocities in the correct places. Incorrect substitution is very common — it is **wrong physics** and you will lose marks.

Example

A car accelerates from rest to $17 \, \text{m s}^{-1}$ in 28 s.

a) How many figures should there be in the statement of the final answer? Explain your answer.
There should be two significant figures in the statement of the final answer because the data in the question have two significant figures.

b) Calculate the acceleration of the car.

$u = 0 \, \text{m s}^{-1}$
$v = 17 \, \text{m s}^{-1}$
$t = 28 \, \text{s}$
$a = ?$

$$a = \frac{v - u}{t}$$
$$\Rightarrow a = \frac{17 - 0}{28} = 0·607 \, \text{m s}^{-2}$$
Acceleration of car = $0·61 \, \text{m s}^{-2}$

Hints & tips

For an object with uniform acceleration from an initial velocity u to final velocity v, average velocity is $\bar{v} = \frac{u + v}{2}$. This is true for all values of u and v and is independent of time t.

Experiment: Measuring acceleration

To measure the acceleration of an object, you have to measure two velocities, *u* and *v*. You also need to measure the time, *t*, for the velocity of the object to change from *u* to *v*. To measure each velocity, you need to measure a displacement and a time.

So, to measure the acceleration of an object, you need to make three time measurements while the object is moving. One of the best ways to do this is to use a computer as shown in Figure 5.2

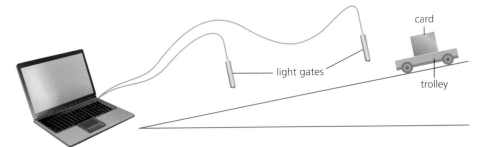

Figure 5.2

- Measure the width of a card and attach the card to the trolley.
- Program the computer to measure the time for the card to cut through each of the two light gates and the time for the trolley to travel from the first light gate to the second light gate.
- To find *u*: divide the width of the card by the time for the card to cut the first light gate.
- To find *v*: divide the width of the card by the time for the card to cut the second light gate.
- To find *a*: first work out (*u* − *v*); divide this by the time for the trolley to travel from the first light gate to the second light gate.

Questions ?

1 **a)** Which of the following quantities are scalars?
- velocity
- speed
- distance
- acceleration
- displacement
- time

  b) Which of the above quantities are vectors?

2 **a)** Which of the following quantities have directions?
- mass
- work
- energy
- weight
- force

  b) Which of the above quantities are vectors? Explain your answer.

3 A boy states that when an object is moving in a straight line, the distance and displacement are always equal. Is the boy correct? Give a reason for your answer.

4 A racing car takes 0·080 s to cross the finishing line. The length of the car is 4·0 m.
  a) Calculate the speed of the car as it finishes the race.
  b) Explain whether this is an average speed or an instantaneous speed.

5 In a race on sports day a girl starts at the finish line and runs two 400-metre laps of the track in a time of 2 minutes and 8 seconds.
  a) Calculate her average speed for the race.
  b) Calculate her average velocity for the race.

6 A boy walks to school in the morning and runs home from school in the afternoon. The distance between the boy's home and the school is 360 m. The boy takes 180 s to walk to school and 120 s to run home. Calculate:
  a) the average speed of the boy on his way to school
  b) the average speed of the boy on his way home.

7 The average speed of a car is 25 m s^{-1}. Calculate the distance travelled by the car in 30 minutes.

8 The average speed of a yacht is 3·8 m s^{-1}. The total distance travelled by the yacht is 11·4 km. Calculate the total time for the yacht to travel this distance.

9 A girl is running 800 m as part of her training for a race. She wants to have an average speed of 5·0 m s^{-1} for the whole run. The girl runs the first 400 m in a time of 160 s. Explain whether the girl can reach an average speed of 5·0 m s^{-1} for the whole run.

10 A boy kicks a ball towards a target. A group of students is asked to measure the instantaneous speed of the ball just after it has been kicked. Another group of students is asked to measure the average speed of the ball from the moment it is kicked until it hits the target.
  a) Describe a method that the first group of students can use to measure the instantaneous speed of the ball.
  b) Describe a method that the second group of students can use to measure the average speed of the ball.
  c) Explain whether the instantaneous speed measured by the first group of students is likely to be greater than, equal to or less than the average speed measured by the second group of students.

11 A car starts from rest and reaches a velocity of 25 m s^{-1} after 10 s. Calculate the average acceleration of the car.

12 A cyclist is travelling along a straight horizontal road at 10 m s^{-1}. The cyclist applies the brakes and comes to rest in a time of 4·0 s. Find the acceleration of the cyclist.

13 At the end of a race a runner enters the final straight running at a velocity of 3·4 m s^{-1}. The runner accelerates at 0·20 m s^{-2} for 3·0 s. Calculate the final velocity of the runner.

14 Apollo 8 had an average acceleration of 25 m s^{-2} during the first 2 minutes of its launch. Calculate the change in velocity of the rocket during this time.

5.2
Adding vectors

What you should know

For N5 Physics you need to be able to:
* ★ understand and use the term resultant
* ★ calculate the resultant of two vectors along a straight line or at right angles
* ★ solve problems on vectors using scale diagrams or calculations.

Key point

* ✱ The sum of two or more vectors is called the **resultant**.

All vectors can be represented by a straight line with an arrow. The length of the line represents the size of the vector. The direction of the arrow shows the direction of the vector.

Adding vectors in a straight line

When adding vectors in a straight line, choose one direction as positive – the other direction is then negative. You may choose either direction as positive (see the Example below). Write down your choice of positive direction – this will help you to use it consistently. Vectors in a straight line are added using normal arithmetic.

Example

An aircraft is flying due east into a wind of $25\,\mathrm{m\,s^{-1}}$ blowing due west. The velocity of the aircraft through the air is $100\,\mathrm{m\,s^{-1}}$. Calculate the resultant velocity of the aircraft.

If you choose the direction of the aircraft as positive:

$\Rightarrow$ velocity of aircraft $= +100\,\mathrm{m\,s^{-1}}$ and velocity of the air $= -25\,\mathrm{m\,s^{-1}}$
$\Rightarrow$ resultant velocity of aircraft $= +75\,\mathrm{m\,s^{-1}}$

(*When the final answer is positive you do not need to include the plus sign to get full marks.*)

If you choose the direction of the wind as positive:

$\Rightarrow$ velocity of aircraft $= -100\,\mathrm{m\,s^{-1}}$ and velocity of the air $= +25\,\mathrm{m\,s^{-1}}$
$\Rightarrow$ resultant velocity of aircraft $= -75\,\mathrm{m\,s^{-1}}$

(*When the final answer is negative you* **must include the minus sign** *to get full marks.*)

Adding vectors at right angles

You can add two vectors at right angles to each other either by calculation or by scale drawing.

By calculation

Use Pythagoras' theorem to get the length of the resultant and the definition of tangent to find the angle – both are needed for your statement of the final answer.

Example

A man walks 120 m due east and then 50 m due north (Figure 5.3). Find his resultant displacement.

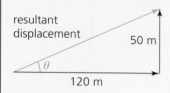

Figure 5.3

Using Pythagoras' theorem: resultant^2 = $120^2 + 50^2 = 16\,900$

$$\Rightarrow \text{resultant} = 130 \,\text{m}$$

$$\tan\theta = \frac{50}{120} = 0.417 \Rightarrow \theta = 22.6°$$

$\Rightarrow$ resultant displacement = 130 m 23° north of east

(In the data '120 m' has three significant figures and '50 m' has two significant figures.

$\Rightarrow$ *minimum number of significant figures in the data is two*

$\Rightarrow$ *two figures are needed in the statement of the final answer.)*

By scale drawing

Vectors are **always** added 'nose to tail'. Choose a scale that gives a big enough drawing but not so big that it is too big for the page. Draw the first vector. Then draw the second vector with its start at the end of the first vector. **Draw carefully**. Make sure that the lengths of your lines and the 90° angle are as accurate as possible.

Example

A girl runs 360 m west and then a further 150 m south (Figure 5.4). Find her resultant displacement.

Scale: 1 cm represents 50 m $\Rightarrow OA = \dfrac{360}{50} = 7{\cdot}2$ cm and $AB = \dfrac{150}{50} = 3{\cdot}0$ cm

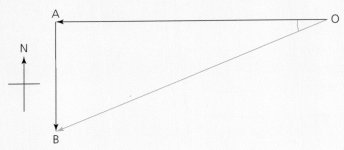

Figure 5.4

Resultant is line OB: by measurement OB = 7·8 cm, which represents 390 m, and angle AOB = 23°

$\Rightarrow$ resultant displacement = 390 m 23° south of west.

Questions

1 A boat sets off at 5·0 m s⁻¹ at right angles to the banks of a river in which the water is flowing at 3·5 m s⁻¹. Find the resultant velocity of the boat. (*Use Pythagoras and tangent.*)

2 A girl runs 300 m west and then 450 m east. She then runs a further 450 m east and finally she runs another 200 m west.
   a) How far did the girl run?
   b) Calculate the girl's resultant displacement.

5.3
Velocity–time graphs

Velocity–time graphs

The motion of an object moving in a straight line can be represented on a velocity–time graph. Most questions on velocity–time graphs include a combination of the shapes below. Learn these shapes – they will help you to describe motions shown in graph questions. Questions may also include upward or downward curves.

Key points

This graph shows a constant positive acceleration.

The initial velocity is $0\,\text{m}\,\text{s}^{-1}$.

When the graph is an upward curve the positive acceleration is not constant – the steeper the curve, the greater the acceleration.

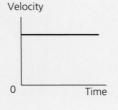

This graph shows a constant velocity.

The acceleration of the object is $0\,\text{m}\,\text{s}^{-2}$.

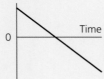

This graph shows a constant negative acceleration.

The initial velocity is positive (above the time axis) and the final velocity is negative (below the time axis).

At the start the object was moving in one direction. It slowed down, stopped and then moved in the opposite direction.

Velocity–time graphs and displacement

The displacement of an object is equal to the area between its velocity–time graph and the time axis. All areas must be counted. Areas above the time axis are positive and areas below the time axis are negative.

Example

The graph in Figure 5.5 represents the motion of a ball thrown into the air.

a) Find the displacement of the ball after 2 s.
   d = area under velocity–time graph
   = area of triangle 1
   = $(\frac{1}{2} \times 2 \times 20)$ = 20 m

b) Find the displacement of the ball after 4 s.
   d = area under velocity–time graph
   = (area of triangle 1) − (area of triangle 2)
   = $(\frac{1}{2} \times 2 \times 20) - (\frac{1}{2} \times 2 \times 20)$ = 0 m

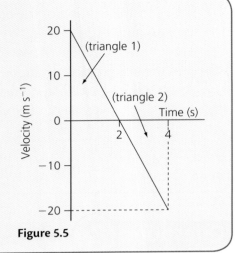

Figure 5.5

Velocity–time graphs and acceleration

The acceleration of an object is equal to the gradient of its velocity–time graph.

Example

The graph in Figure 5.6 represents the motion of a bus travelling between two bus stops.

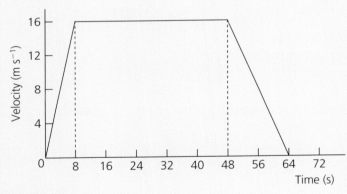

Figure 5.6

a) Find the acceleration of the bus during the first 8 s.
   a = gradient of velocity–time graph
   $= \dfrac{16-0}{8-0}$
   = 2·0 m s^{-2}

b) State the acceleration of the bus at time $t = 30\,s$. Give a reason for your answer.

At $t = 30\,s$, $a = 0\,ms^{-2}$ (Reason: The graph is parallel with the time axis.)

c) Find the acceleration of the bus at time $t = 60\,s$.

At $t = 60\,s$, a = acceleration between $t = 48\,s$ and $t = 64\,s$

= gradient of velocity–time graph

$$= \frac{0 - 16}{64 - 48}$$

$$= -1 \cdot 0\,ms^{-2}$$ (*This acceleration is negative.*)

Questions ?

1 A velocity–time graph is shown in Figure 5.7. Look carefully at the graph and answer the questions that follow.

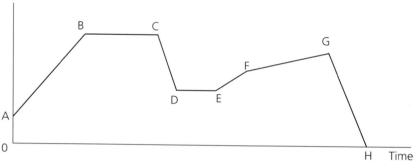

Figure 5.7

a) Is the initial velocity of the object $0\,ms^{-1}$? Explain your answer.

b) Between which pair or pairs of points is the velocity of the object:

i) increasing   ii) decreasing   iii) constant?

c) State the final velocity of the object.

2 The velocity–time graph of an object is shown in Figure 5.8. Look carefully at the graph and answer the questions that follow.

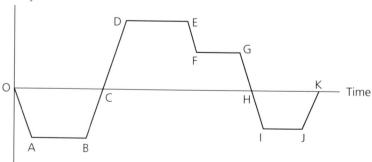

Figure 5.8

a) Between which pair or pairs of points is the velocity of the object:
   i) positive   ii) negative?
b) At which points is the velocity of the object $0\,\mathrm{m\,s^{-1}}$?
c) Between which pair or pairs of points is the acceleration of the object:
   i) positive   ii) negative   iii) zero?

3 A car, travelling on a long straight road, starts from rest and accelerates to a velocity of $20\,\mathrm{m\,s^{-1}}$ in a time of 15 s. The car travels at this velocity for 45 s. The driver notices a friend at the roadside and applies the brakes for 20 s, bringing the car to a stop some distance past his friend. The driver keeps the car stationary for 5 s, then reverses back, reaching a velocity of $5\,\mathrm{m\,s^{-1}}$ after 5 s. He stays at this velocity for 10 s, and brings the car to rest in a further 5 s.
   Draw the velocity–time graph for the car.

4 The graph in Figure 5.9 represents the speed of a cyclist on a cycle track.

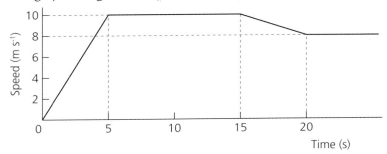

Figure 5.9

a) Calculate the distance travelled by the cyclist in the first 5 s.
b) Calculate the distance travelled by the cyclist in the first 15 s.
c) i) How far does the cyclist travel between $t = 15\,\mathrm{s}$ and $t = 20\,\mathrm{s}$?
   ii) Give a reason for the cyclist slowing down between these times.

5 The graph in Figure 5.10 represents the motion of a bouncing ball.

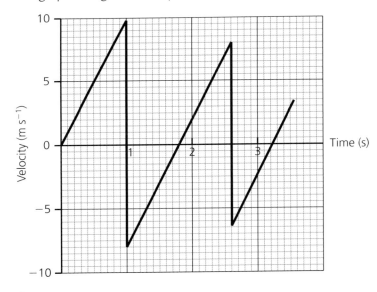

Figure 5.10

a) Describe the motion of the ball during the first 2 s.
b) Calculate the initial height of the ball.
c) Calculate the initial acceleration of the ball.
d) Calculate the height of the first bounce of the ball.
e) In this question, which direction is positive? Explain your answer.

6 A scientist is testing the performance of two cars, A and B.
   a) In an engine test, car A accelerates from $0.0\,\mathrm{m\,s^{-1}}$ to $30\,\mathrm{m\,s^{-1}}$ in 7.9 s, while car B accelerates from $0.0\,\mathrm{m\,s^{-1}}$ to $25\,\mathrm{m\,s^{-1}}$ in 7.2 s.
   Which car has the greater average acceleration? Justify your answer by calculation.
   b) In a brake test, it takes 15 s to bring car A to rest from an initial speed of $24\,\mathrm{m\,s^{-1}}$, while it takes 12 s to bring car B to rest from an initial speed of $20\,\mathrm{m\,s^{-1}}$.
      i) On the basis of this test, which car has the more effective brakes? Justify your answer by calculation.
      ii) This is not a fair test of the braking systems. How should the test be changed to make it fair?

5.4
Force, gravity and friction

What you should know

For N5 Physics you need to be able to:

★ understand and use the terms force, mass, weight and gravitational field strength
★ describe the effects of forces
★ understand and use the relationship $W = mg$
★ understand and describe the effects of friction on moving objects.

Forces are interactions between the materials of our Universe. We cannot see forces but we can observe the effects of forces.

Force is a vector. The SI unit of force is the **newton** (N) and the symbol for force used by SQA is F. A force of 1 N will accelerate a 1 kg mass at $1\,\mathrm{m\,s^{-2}}$.

(A force of 1 N is about the weight of an average-sized apple.)

Key points

Effects of forces: forces can change the:

* shape
* speed
* direction of an object.

When you observe any or all of these effects then you know a force or forces are acting.

Mass and weight

Mass is a fundamental property of all material in the Universe. It is a measure of the amount of material in objects. The mass of an object is constant – it does not change when the object moves, even if the object goes into space or to another planet.

Mass is a scalar. The SI unit of mass is the **kilogram** (kg) and the symbol for mass used by SQA is m.

Weight is the gravitational force acting on an object. It depends on the mass of the object and the strength of the gravitational field.

Weight is a vector. The SI unit of weight is the **newton** (N) and the symbol for weight used by SQA is W.

On Earth, the weight of an object is the pull of planet Earth on the object. The weight of an object can and does change as it moves. For example, the weight of an object on Mars is different from the weight of the same object on Jupiter or on Earth or in outer space – all of these weights are different.

Gravitational force

Gravitational force is a force of **attraction** between objects that have mass.

Gravitational field strength is the gravitational force acting on a unit mass. It is a vector. The SI unit for gravitational field strength is **newton per kilogram** ($N\,kg^{-1}$) and the symbol for gravitational field strength used by SQA is g. The relationship between mass, weight and gravitational field strength is $W = mg$.

On Earth the gravitational field strength is approximately $9.8\,N\,kg^{-1}$.

Hints & tips

If you are asked a question about the weight of objects or people on the moon, the Earth or any other planet, you will be given the gravitational field strengths in the Data Sheet provided.

Gravitational field strength is numerically equal to acceleration due to gravity **when frictional forces can be ignored**. SQA uses the same symbol, g, for both quantities.

Example

An astronaut has a weight of 608 N on planet Earth. Calculate her weight on the moon. (Gravitational field strength of the moon = $1.6\,N\,kg^{-1}$.)

On Earth, $\quad W = 608\,N \qquad\qquad\qquad W = mg$

$\qquad\qquad g = 9.8\,N\,kg^{-1} \qquad \Rightarrow \quad 608 = m \times 9.8$

$\qquad\qquad\qquad\qquad\qquad\qquad\quad \Rightarrow \quad m = 62.0 = 62\,kg$

On the moon, $m = 62\,kg \qquad\qquad\quad W = mg$

$\qquad\qquad g = 1.6\,N\,kg^{-1} \qquad\qquad\qquad = 62 \times 1.6$

$\qquad\qquad\qquad\qquad\qquad\qquad\qquad\qquad = 99.2$

$\qquad\qquad\qquad\qquad\qquad\qquad\qquad\qquad = 99\,N$

Why is the final answer rounded to two figures? (*Think* – **significant** *figures!*)

The force of friction

Frictional forces occur when materials are in contact. The size of the force depends on the materials. For example, when you walk the force of friction between your shoes and the pavement stops you from slipping. When the pavement is covered with ice the force of friction between your shoes and the ice is much smaller and you are more likely to slip.

Friction can oppose the motion of objects. This effect may be different for different objects. For example, think of the difference between the motion of a falling hammer and a falling feather. Air resistance has a much greater effect on the feather.

Sometimes the force of friction is helpful so we try to make it bigger. For example, when a car is going around a corner, the force of friction between the tyres and the road helps to keep the car on the road. A fully laden lorry going around the same corner at the same speed needs a much bigger force to keep it on the road, so its tyres are designed to make the force of friction large.

At other times the force of friction is not helpful and so we try to make it smaller. For example, a skier in a downhill race wants to get to the finish line as quickly as possible. The skier tries to make the force of friction as small as possible by waxing the underside of the skis.

Example

A 45 kg child on a slide is at a height of 3·0 m.
He slides down and his speed at the bottom is $5{\cdot}0\,\text{m s}^{-1}$
How much energy is 'lost' to friction?

$m = 45$ kg $\qquad E_p = mgh$
$g = 9{\cdot}8$ N kg^{-1} $\qquad = 45 \times 9{\cdot}8 \times 3{\cdot}0 = 1323$ J
$h = 3{\cdot}0$ m $\qquad E_k = \frac{1}{2}mv^2$
$v = 5{\cdot}0$ m s^{-1} $\qquad = 0{\cdot}5 \times 45 \times 5{\cdot}0^2 = 562{\cdot}5$ J

Energy lost to friction $= 1323 - 562{\cdot}5 = 760{\cdot}5$ J $= 761$ J
This energy is converted mainly into heat as the child slides down.

Questions ?

1 a) Describe two examples of force changing the shape of an object.
   b) Describe two examples of force changing the speed of an object.
   c) Describe two examples of force changing the direction of motion of an object.
2 The mass of a car is 1220 kg. Calculate the weight of the car (on planet Earth).
3 A surface vehicle of mass 2130 kg is sent to collect samples of rocks from the surface of Mars. Calculate the change in the weight of the surface vehicle as it travels from Earth to Mars.
   Gravitational field strength on Mars $= 3{\cdot}7$ N kg^{-1}
   Gravitational field strength on Earth $= 9{\cdot}8$ N kg^{-1}
4 a) Describe one situation where the force of friction is not helpful. Describe one way of reducing this frictional force.
   b) Describe one situation where the force of friction is helpful. Describe one way of increasing this frictional force.

5.5
Newton's Laws

What you should know

For N5 Physics you need to be able to:

★ understand and use the terms balanced forces, unbalanced forces, resultant force and reaction force
★ understand and use Newton's Laws of Motion
★ explain the movement of an object in terms of the resultant force acting on the object
★ use diagrams to analyse forces acting on an object
★ find the resultant of two forces acting at right angles to each other
★ understand and use the relationship $F = ma$.

Balanced forces

When you sit on a chair there are at least two forces acting on you – the force of gravity downwards and the force from the chair upwards. These two forces are equal in size and opposite in direction. They both act on you **and** they act in the same line so they cancel each other out. There is no resultant force and we say these forces are **balanced**.

Key points

* When the forces acting on an object are balanced, the motion of the object stays the same. This is **Newton's First Law of Motion**.
* It means that a stationary object stays stationary.
* A moving object does not get faster or slower or change direction – it travels at a constant speed in a straight line.

Unbalanced forces

When you stand up, the upward force from your muscles is greater than the downward force of gravity. These forces do not cancel each other out. There is a resultant force that makes you change speed. We say these forces are **unbalanced**.

Example

For each of the following, explain whether the forces acting are balanced or unbalanced:

a) a bus stationary at a bus stop
   The bus is stationary – its motion is not changing so the forces are balanced.

b) a bus turning a corner

The bus is turning a corner – the direction of its motion is changing so the forces acting on the bus are unbalanced.

c) a car braking to a stop.

The speed of the car is changing so the forces acting on the car are unbalanced.

When the forces acting on an object are unbalanced, the object **accelerates** – that is, it gets faster **or** slower **or** changes direction. A constant unbalanced force produces a constant acceleration.

The direction of the acceleration is the same as the direction of the unbalanced force.

Key points

* When an unbalanced force acts on an object, the acceleration is directly proportional to the unbalanced force and inversely proportional to the mass of the object. This is **Newton's Second Law of Motion**.
* The relationship between resultant force, mass and acceleration is $F = ma$.

Hints & tips

When you are asked to describe how changing mass or force affects the acceleration of an object, use the relationship in the form $a = \dfrac{F}{m}$ and consider how the value of $\dfrac{F}{m}$ changes.

For example, when force is increased and mass stays the same, the value of a increases. When mass is increased and force stays the same, the value of a decreases.

In many N5 Physics questions you will be given the value of the resultant force. In other questions you will have to work out the value of the resultant force before you use the relationship. (In some questions, resultant force is simply called 'force'.)

Example

A space rover sent to Mars has a mass of 900 kg. Calculate the minimum engine force required to ensure that the rover has a soft landing on Mars.

Minimum engine force = weight of the rover on Mars

$m = 900$ kg

$g = 3.7$ N kg^{-1}

$W = ?$

$W = mg$

$= 900 \times 3.7$

$= 3330$ N

$\Rightarrow$ minimum engine force $= 3.3$ kN

Reaction forces

When you are walking along the road you push backwards with the muscles in your legs. You move forward because your push on the ground causes a reaction force on you in the opposite direction. This reaction force is exactly the same size as your push.

Key points

* Every force causes an equal and opposite force. This is **Newton's Third Law of Motion**.
* When two objects interact, the force between them is the same in both directions. For example, a girl standing on a floor exerts a downward force on the floor. The floor exerts an equal and opposite force on the girl.
* Action-reaction forces occur between two different objects. Each force acts on one of the objects.

Example ⚑

For each of the following forces, identify its reaction force:

a) Force exerted on a rocket by the fuel.
b) Air resistance acting on an aeroplane in flight.
c) Weight of a car.

Reaction forces are:

a) Force exerted on the fuel by the rocket.
b) Force exerted by the aeroplane on the air.
c) Careful with this one! Weight is the force of gravity exerted by the Earth on the car. The reaction force is the force of gravity exerted by the car on the Earth.

Analysing forces acting on objects

The first step is to name the forces and identify the directions in which they act. For example, when a plane is flying there are four forces acting. Figure 5.11 shows these forces and the directions in which they act.

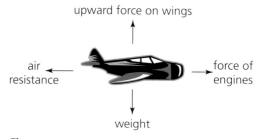

upward force on wings

air resistance

force of engines

weight

Figure 5.11

You are unlikely to be asked a question about a situation more complicated than this.

The second step is to consider the sizes of the forces acting on the object. The motion of the object depends on the sizes of these forces and whether they are balanced or unbalanced. For example, when the plane is flying horizontally at a constant speed, the forces are balanced. If the pilot increases the engine force, the speed of the plane increases.

Example

A plane is flying horizontally at constant speed. The pilot reduces the engine force. Explain why the speed of the plane decreases.

Plane flying horizontally at constant speed $\Rightarrow F_{engine} = F_{air\ resistance}$

Force of engine reduced $\qquad\qquad \Rightarrow F_{engine} < F_{air\ resistance}$

The forces are now unbalanced. The unbalanced force causes the plane to slow down.

Adding forces in a straight line

Remember that force is a vector. Forces acting in opposite directions have opposite signs – one direction is positive, the other direction is negative.

Example

Four men start to push a car up a slope. The weight of the car acting down the slope is 785 N. To get the car moving, the men push with forces of 220 N, 185 N, 215 N and 225 N.

a) Calculate the resultant force acting on the car.

Total force up the slope $\quad = 220 + 185 + 215 + 225 = 845\,N$

Total force down the slope $\quad = -785\,N$ *(Note the minus sign!)*

$\Rightarrow$ Resultant force $\qquad\quad = 845 - 785 = 60\,N$

b) Apart from the weight of the car, what other force will the men have to overcome to push the car up the slope?

The men will also have to overcome frictional forces between the tyres and the road.

Adding forces at right angles

To add forces at right angles to each other you can do a scale drawing or use Pythagoras and tangent.

Example ⚑

In an experiment, three students are using newton balances to investigate the resultant of two forces. Students A and B use newton balances to apply forces of 3·6 N and 2·7 N at right angles to each other, as shown in Figure 5.12.

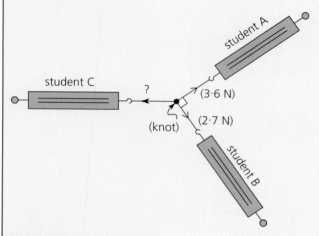

Figure 5.12

Student C holds the third newton balance steady.

a) By using a scale drawing (such as Figure 5.13), find the resultant of the forces applied by students A and B.

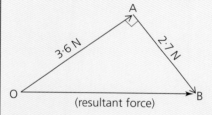

Figure 5.13

> Scale: 1 cm represent 1·0 N. (OA = 3·6 cm, angle OAB = 90°, AB = 2·7 cm)
> Resultant is OB, OB = 4·5 cm, angle BOA = 37°.
> Hence resultant force = 4·5 N at 37° to the 3·6 N force.

b) State the size of the reading on the newton balance held by student C. Explain your answer.
> Reading on newton balance held by student C = 4·5 N. The force exerted by the balance held by student C is equal and opposite to the resultant of the two forces applied by students A and B.

Questions ?

1 For each of the following explain whether the forces acting are balanced or unbalanced:
   a) a girl running at a constant speed in a straight line
   b) a cyclist moving at a constant speed around a corner
   c) an aeroplane starting its take-off
   d) a car stopping at traffic lights
   e) an electron at rest in an electric field
   f) an electron moving at constant speed in a circle.

2 a) Give two examples of situations where the forces acting on an object are balanced.
   b) Give two examples of situations where the forces acting on an object are unbalanced.

3 For each of the following, give the names and directions of the forces acting on the object.
   a) Racing car accelerating at the beginning of a straight part of the track.
   b) Family car slowing before it reaches a bend in the road.
   c) i) State whether the forces in part a) are balanced or unbalanced.
      ii) State whether the forces in part b) are balanced or unbalanced.

4 A train is travelling along a straight horizontal track. The engine of the train exerts a forward force of 8300 N. A total frictional force of 3200 N acts on the train.
   a) Calculate the resultant force acting on the train.
   b) The total mass of the train is 3.4×10^4 kg. Calculate the acceleration of the train.

5 Two forces of 500 N are applied, as shown in Figure 5.14, to drag a heavy crate across a factory floor. The crate moves at a constant speed of $0.10\,\mathrm{m\,s^{-1}}$ in the direction shown. Calculate the force of friction between the base of the crate and the factory floor.

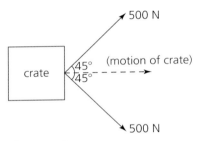

Figure 5.14

6 A car of mass 1200 kg has a uniform acceleration of $0.70\,\mathrm{m\,s^{-2}}$. Calculate the unbalanced force acting on the car.

7 A trolley accelerates uniformly at $500\,\mathrm{mm\,s^{-2}}$ when an unbalanced force of 2.0 N is acting on the trolley. Calculate the mass of the trolley.

8 A cyclist moving at $2.4\,\mathrm{m\,s^{-1}}$ pedals harder for 5.0 s and reaches a speed of $6.4\,\mathrm{m\,s^{-1}}$. The combined mass of the cyclist and bicycle is 80 kg. Calculate the average unbalanced force acting while the cyclist is accelerating.

9 A car manufacturer is marketing a model with an engine that produces a force 5% less than normal. The reduced power model is 10% cheaper. The mass of the reduced power model is 2% less than the normal car. How does the acceleration of the lower-power model compare with the acceleration of the normal car?

10 Action-reaction forces are equal in size and opposite in direction and yet they do not cancel each other out. Explain.

5.6
Energy, work and power

What you should know

For N5 Physics you need to be able to:

★ define and use the terms energy, work, power and efficiency
★ state and use the principle of conservation of energy
★ understand and use these relationships to solve numerical problems

$$E_w = Fd, \ P = \frac{E_w}{t}, \ E_p = mgh, \ E_k = \frac{1}{2}mv^2$$

$$\text{percentage efficiency} = \frac{\text{useful } E_o}{E_i} \times 100$$

Conservation of energy

When an object loses energy, the lost energy is converted to other forms of energy. When an object gains energy, the energy gained has been converted from other forms of energy. This is true for all energy changes.

Key points

✳ Energy cannot be created and it cannot be destroyed.
✳ Energy is a scalar. The SI unit of energy is the **joule** (J) and the symbol for energy used by SQA is E. (The symbol E often has a subscript to identify the particular type of energy, for example E_k for kinetic energy.)

Work

When an unbalanced force F moves through a distance d, work is done and energy is converted from one form to another.

Work is a scalar. The SI unit of work is the **joule** (J) and the symbol for work used by SQA is E_W.

The relationship between work, force and distance is $E_W = Fd$.

One joule of work is done when a force of 1 N moves through a distance of 1 m.

Power

Power is the rate at which work is done or the rate at which energy is converted from one form to another.

Power is a scalar. The SI unit of power is the **watt** (W) and the symbol for power used by SQA is P. (The symbol P often has subscripts to differentiate power out from power in.)

The relationship between power, work and time is $P = \dfrac{E_w}{t}$.

1 watt is equal to 1 **joule per second** (Js^{-1}).

Example ⚑

A man exerts a force of 200 N for 6 minutes in pushing his car 60 m along a horizontal road to reach a garage.

a) Calculate the work done by the man.

$E_w = ?$ $\qquad\qquad\qquad E_w = Fd$

$F = 200\,N$ $\qquad\qquad\qquad\quad = 200 \times 60$

$d = 60\,m$ $\qquad\qquad\qquad\quad = 12\,000\,J$

b) Calculate the power of the man when he was pushing the car.

$P = ?$ $\qquad\qquad\qquad\qquad P = \dfrac{E_w}{t}$

$E_w = 12\,000\,J$ $\qquad\qquad\qquad = \dfrac{12\,000}{360}$

$t = 6\,min = 360\,s$ $\qquad\qquad = 33\cdot3$

$\qquad\qquad\qquad\qquad\qquad\quad = 33\,W$

Gravitational potential energy

Gravitational potential energy is gained or lost by an object as it moves **vertically**. Gravitational potential energy is a scalar. The SI unit of gravitational potential energy is the **joule** (J) and the symbol for gravitational potential energy used by SQA is E_p.

The relationship for the potential energy of an object of mass m, at height h, in gravitational field strength g, is $E_p = mgh$.

In some N5 questions, 'gravitational potential energy' may be shortened to 'potential energy'.

Kinetic energy

Kinetic energy is energy that an object has because of its mass and its motion. Kinetic energy is a scalar. The SI unit of kinetic energy is the **joule** (J) and the symbol for kinetic energy used by SQA is E_k.

The relationship for the kinetic energy of an object of mass m and velocity v is $E_k = \frac{1}{2}mv^2$.

Example

A boy of mass 45 kg is standing at a height of 2·1 m on a climbing frame. The boy jumps to the ground.

a) Calculate the potential energy lost by the boy during his jump.

$E_p = ?$ $E_p = mgh$

$m = 45\,kg$ $= 45 \times 9{\cdot}8 \times 2{\cdot}1$

$h = 2{\cdot}1\,m$ $= 926{\cdot}1\,J$

$g = 9{\cdot}8\,N\,kg^{-1}$ $= 930\,J$

b) Hence calculate the speed of the boy when he lands.

$v = ?$ $E_k = \frac{1}{2}mv^2$

$E_k = 926{\cdot}1\,J$ $\Rightarrow$ $926{\cdot}1 = 0{\cdot}5 \times 45 \times v^2$ *(Note use of the un-rounded value of E.)*

$m = 45\,kg$ $\Rightarrow$ $v = 6{\cdot}41 = 6{\cdot}4\,m\,s^{-1}$

c) State any assumption you make to solve part b).

Assumption – all of the potential energy is converted to kinetic energy.

Key points !

potential energy (E_p) = mass × gravitational field strength × height

$$E_p = mgh$$

Kinetic energy (E_k) = $\frac{1}{2}$ × mass × velocity^2

$$E_k = \frac{1}{2}mv^2$$

Efficiency

All machines and processes waste energy. This means that the energy put in (input energy), E_i, is more than the useful energy that comes out (output energy), E_o.

Efficiency is a measure of how well energy is used. Efficiency is usually expressed as a percentage.

One relationship for calculating efficiency is:

$$\text{percentage efficiency} = \frac{useful\ E_o}{E_i} \times 100$$

Efficiency has no unit – you will lose marks if you include a unit in the statement of your final answer when you calculate efficiency.

Efficiency may also be calculated from power, using the relationship:

$$\text{percentage efficiency} = \frac{useful\ P_o}{P_i} \times 100$$

Example

The electric motor of a crane uses 42 000 J of electrical energy in lifting a pack of eight 25 kg bags of cement through a distance of 15 rn from the ground to the fourth floor of a block of flats. Calculate the efficiency of the motor during the lifting process.

$$useful\ E_o = E_p = ?$$

$$m_{cement} = 8 \times 25 = 200\,kg$$

$$g = 9{\cdot}8\,N\,kg^{-1}$$

$$h = 15\,m$$

$$percentage\ efficiency = ?$$

$$E_i = 42\,000\,J$$

$$E_p = mgh$$

$$= 200 \times 9{\cdot}8 \times 15$$

$$= 29\,400\,J$$

$$\Rightarrow\quad useful\ E_o = 29\,400\,J$$

$$percentage\ efficiency = \frac{useful\ E_o}{E_i} \times 100$$

$$= \frac{29\,400}{42\,000} \times 100$$

$$= 70\%$$

Questions

1 A force of 30 N is used to move a crate across a factory floor. The energy used moving the crate is 450 J. Calculate the distance moved by the crate.

2 A pendulum consists of a small metal sphere suspended at the end of a long string. The metal sphere is pulled to the side and released from a point 35 mm higher than the lowest part of its swing. Calculate the maximum velocity of the metal sphere.

3 A power boat of mass 510 kg travels at a constant speed of 6·0 m s^{-1} for 15 minutes. The input power of the boat engine is 12 000 W. The propeller provides a forward force of 850 N.
   a) Calculate the energy used by the boat engine.
   b) Calculate the distance travelled by the boat.
   c) Calculate the useful energy output of the boat engine.
   d) The answer to part c) is much smaller than the answer to part a). Explain this in terms of forces acting on the boat.
   e) Energy cannot be created or destroyed – so where has the 'missing' energy gone?

4 Percentage efficiency has no unit. Explain.

5 Calculate the speed with which a 0·7 kg ball hits the ground when dropped from 3·0 m.

6 Calculate the speed with which a 1·5 kg ball hits the ground when dropped from 3·0 m.

7 A rock is thrown upwards with a velocity of 6·0 m s^{-1}. Calculate the maximum height it could reach.

8 A pendulum's bob is raised 0·45 m above its rest position. What velocity will it reach at its lowest point?

9 Calculate the maximum velocity an object will reach when dropped from:
   a) 1·0 m   b) 2·0 m   c) 3·0 m
   d) 4·0 m   e) 5·0 m

10 Draw a graph of velocity against height for the object being dropped in question 9.

11 A 0·45 kg toy slides down a slope. It travels a vertical distance of 1·2 m. Its speed at the bottom is 3·5 m s^{-1}.
   a) Calculate its gravitational potential energy at the top of the slope.
   b) Calculate the energy 'lost' due to friction.
   c) What would the maximum possible speed be if we could reduce friction to zero?

12 A grandfather clock operates by raising a mass and allowing it to slowly fall. A 2·0 kg mass is raised to a height of 1·5 m.
   a) Calculate the gravitational potential energy gained by the mass as it is raised.
   b) The clock uses 0·083 joules per hour. How long can it run for before the mass needs to be raised again?
   c) Into what is the potential energy transformed?

13 A 1·1 kg ball is dropped from a height of 1·5 m. It rebounds to a height of 1·3 m.
   How much energy was transformed into heat and sound during this?

14 A pendulum is raised and released. Explain why it will eventually slow down and stop.

15 A car runs out of petrol while driving. Why does it slow down and stop?

16 Cruachan Power Station is described as a pumped storage scheme. Explain why water is pumped up to the reservoir at times of low demand.

5.7
Projectiles

What you should know

For N5 Physics you need to be able to:

★ explain the shape of the path of a projectile
★ use relationships and graphs to solve numerical problems on projectiles
★ explain terminal velocity
★ explain satellite orbits in terms of projectile motion.

Key point

✳ **Projectiles** are objects that are moving in an area of space where there is gravitational force.

Projectiles with vertical motion only

When an object is dropped vertically or thrown vertically upward, the motion of the object is along a straight line. When air resistance can be ignored, downward acceleration of the object is constant. On Earth, **when frictional forces are negligible**, the vertical acceleration of a projectile is $9.8\,\mathrm{m\,s^{-2}}$.

Example

A tennis ball is thrown vertically upwards with a velocity of $4.9\,\mathrm{m\,s^{-1}}$. Calculate the time taken for the ball to return to its starting position. Air resistance is negligible.

$u = +4.9\,\mathrm{m\,s^{-1}}$ (*Choosing u as positive means that 'up' is the positive direction.*)

$v = -4.9\,\mathrm{m\,s^{-1}}$ (*It returns to the same position at the same speed.*)

$a = -9.8\,\mathrm{m\,s^{-2}}$ (*Acceleration is negative as it is down.*)

$$a = \frac{v - u}{t}$$

$$\Rightarrow \quad -9.8 = \frac{-4.9 - 4.9}{t}$$

$$\Rightarrow \quad t = 1.0\,\mathrm{s}$$

Terminal velocity

On Earth, as a falling object gets faster, the force of air resistance increases. If an object falls far enough, the air resistance becomes equal to the weight of the object. The object then moves with a constant downward velocity, which is called its **terminal velocity**. The size of an object's terminal velocity depends on the mass and shape of the object. It also depends on the material through which it is falling.

A falling object can reach a terminal velocity **only** if it is falling through a gas or a liquid.

Projectiles with both horizontal and vertical motion

An object that is projected horizontally has both vertical motion and horizontal motion. Its horizontal motion has no effect on its vertical motion **and** its vertical motion has no effect on its horizontal motion.

Example

A football kicked horizontally from a cliff has a vertical velocity of $29.4\,\text{m s}^{-1}$ when it lands in the sea below.

a) Calculate the time the ball takes to fall.

$$u = 0\,\text{m s}^{-1} \qquad\qquad a = \frac{v - u}{t}$$

$$v = 29.4\,\text{m s}^{-1} \qquad \Rightarrow \quad 9.8 = \frac{29.4 - 0}{t}$$

$$a = 9.8\,\text{m s}^{-2} \qquad \Rightarrow \quad t = 3.0\,\text{s}$$

b) The initial horizontal velocity of the ball is $15\,\text{m s}^{-1}$. Calculate the horizontal distance travelled by the ball (Figure 5.15).

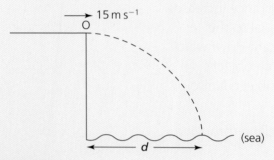

Figure 5.15

$$v_{\text{horizontal}} = 15\,\text{m s}^{-1} \qquad\qquad d = vt$$

$$t = 3.0\,\text{s} \qquad\qquad\qquad = 15 \times 3.0 = 45\,\text{m}$$

Satellites

A satellite is a special projectile. The path followed by a satellite is called its orbit. Satellites travel around a much larger and more massive object in a circular or elliptical orbit (an ellipse is a slightly flattened circle).

A satellite in a circular orbit has a large, constant horizontal speed that is changing direction all the time. These satellites are constantly falling without getting closer to the object they orbit. Puzzled? Consider Figure 5.16, which shows part of an orbit of a satellite of the Earth.

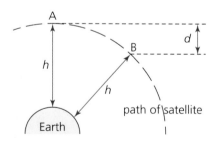

Figure 5.16 Part of a satellite's orbit around Earth

The satellite moves from point A to point B. At A the satellite is at a height h above the Earth. The satellite fell a distance d as it moved to point B. The surface of the Earth has curved away from the satellite, so at B the object is still at a height h above the Earth.

Example

Are the forces acting on a satellite balanced or unbalanced? Explain your answer.

There is an unbalanced force acting on a satellite. The direction of the satellite's motion is changing all the time so there must be an unbalanced force acting all the time.

Questions

In the questions that follow you may use the following data: $g = 9.8\,\text{m s}^{-2}$.

1 A stone dropped down a well takes 3·6 s to reach the surface of the water. How far is the water surface below the top of the well?

2 A rocket fired vertically upwards with a velocity of 60 m s^{-1} falls back to Earth. Ignoring the effects of air resistance, calculate the height reached by the rocket.

3 An object that is projected horizontally follows a curved path. Explain the shape of the path.

4 A ball bearing, projected horizontally with a velocity of 2·0 m s^{-1}, hits the ground after 0·25 seconds.
   a) Draw the velocity–time graph of the horizontal motion of the ball bearing. Your graph must indicate the velocity at $t = 0.25$ s. (You may find it useful to refer back to Topic 5·3, Velocity–time graphs.)
   b) Draw the velocity–time graph of the vertical motion of the ball bearing. Your graph must indicate the velocity at $t = 0.25$ s.
   c) Calculate the horizontal displacement of the ball bearing as it hits the ground.
   d) Calculate the vertical displacement of the ball bearing as it hits the ground.

5 A bird flying horizontally at 4·8 m s^{-1} drops a stone from its beak. The stone hits the ground after it has travelled a horizontal distance of 12 m.
   a) After the bird dropped it, how long did it take the stone to fall to the ground?
   b) Calculate the vertical velocity of the stone when it hits the ground.

6 At the start of a game of tennis, a girl hits a tennis ball so that its initial velocity is horizontal. The ball follows a curved path and bounces 1·5 s later.
   Using only the information given in the question and the value of g, is it possible to calculate the horizontal distance travelled by the tennis ball? Explain your answer.

7 A parachutist jumps from an aeroplane. In terms of the forces acting on her, explain how her acceleration changes as she falls.

8 A feather and a hammer are dropped at the same time from a height of 25 m above the surface of the moon. Which object is first to hit the moon surface? Explain your answer.

   (At the beginning of August 1971, Commander David Scott carried out a similar experiment during the Apollo 15 mission to the moon – he dropped a hammer and a feather from a height of approximately 1·6 m. The experiment was broadcast live and a video of the event is available on a popular internet video-sharing site!)

Chapter 6: Space

Space exploration

What you should know

For N5 Physics you need to be able to:

★ describe our current beliefs about the Universe using evidence gathered from telescopes and satellites
★ discuss the impact of space exploration on our understanding of our own planet
★ consider the benefits to us of associated technologies from space exploration and assess their impact
★ understand the risks and benefits associated with space exploration, including the challenges of re-entry
★ use and convert distances between metres and light years.

Our planet, Earth, is one of many large objects that orbit the Sun. We refer to the Sun and planets as our Solar System, but our Solar System actually extends far beyond the last planet. It is difficult to determine the exact range of the Solar System but it could be as far as 100 times the distance from the Earth to the Sun.

Our Sun is one of many stars that make up our Galaxy, the Milky Way. The Milky Way is approximately 110 000 **light years** in diameter and contains 200–400 billion stars. There are millions of galaxies in the Universe. The distances involved are so vast that we use a special unit, the light year, to describe them.

Key points (!)

One **light year** is the distance travelled by light in one year. That is:

$$3 \times 10^8 \times 60 \times 60 \times 24 \times 365 \cdot 25 = \mathbf{9 \cdot 47 \times 10^{15}\,m}$$

It is important to remember that it is a measure of distance **not** time.

Example

Barnard's Star is 5·96 light years from us. Calculate the distance in km.

$$d = 5 \cdot 96 \times 9 \cdot 47 \times 10^{15}$$
$$= 5 \cdot 64 \times 10^{16}\,m$$
$$= 5 \cdot 64 \times 10^{13}\,km$$

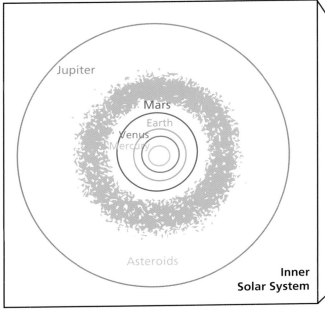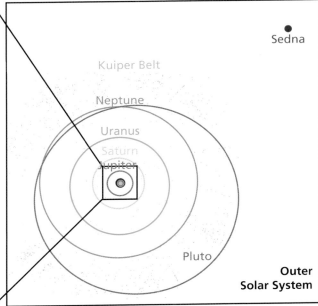

Figure 6.1 The inner and outer Solar System

The Solar System extends far beyond the planets and asteroid belt we are familiar with. It is generally accepted to be the volume of space where the dominant energy source or gravitational effect is from the Sun. Astronomers have observed many objects that are trapped by the Sun's gravity but orbit at much greater distances. Figure 6.1 gives an indication of the wider areas of the Solar System.

Our information regarding the composition and size of the Solar System comes mainly from Earth-based telescopes. In recent years we have been able to augment this information with data from satellite-based telescopes and from probes that we have sent out to the farther reaches of the Solar System.

Pluto is shown as a planet in many old posters and texts, but it was declassified as a planet in 2006. The reason was that there are a number of similarly sized objects in a region of our Solar System known as the Kuiper Belt. If Pluto were to remain as a planet then these other very distant objects would also have be given planet status. Instead, Pluto and these other objects are now known as **dwarf planets**.

Generally, a dwarf planet is (in size terms) somewhere between a planet and a natural satellite. It is large enough for gravity to form it into a nearly spherical shape, but its gravity is not great enough to attract or clear the other material and objects in its orbit. **Asteroids** are generally smaller than dwarf planets and are sometimes referred to as minor planets. Asteroids are similar to dwarf planets, and indeed all planets, in that they are composed of rock-type materials, mainly dust that has come together by gravitational attraction.

Asteroids are found throughout the Solar System, many in the asteroid belt, and they can range from a few metres across to nearly 1000 km. The larger ones can be almost spherical in shape, but the smaller ones are irregular.

We often see images of asteroids closely packed together in the asteroid belt. This is incorrect. The distance between asteroids in the belt is huge. If you were to fly through the belt, your chances of hitting an asteroid would be minimal.

The planets we know about all orbit a central star. As there are billions of stars in the Universe, there will be billions of planets orbiting the stars outside our Solar System.

These are referred to as **exoplanets** and we have discovered thousands of them and that number is rising constantly.

As they are so distant and small, it is almost impossible to detect them directly by observing them with a telescope. We detect them indirectly. This means we cannot observe them, but we can observe and measure the effect they have on nearby bodies.

1 Changes in the velocity of a star

Astronomers can measure the movement of a star very accurately. Some stars have been found to 'wobble' slightly. This can be due to the gravitational pull of an orbiting planet. By measuring the frequency and magnitude of the wobble, we can estimate the mass and orbital period of nearby planets.

2 Changes in the brightness of a star

When a planet passes in front of a star, it blocks some of the light from that star which would have radiated to Earth. Modern telescopes can measure the brightness of stars very accurately and a slight dimming can indicate the presence of a planet. The amount and duration of the dimming can give information about the mass and period of that orbiting planet.

There are other methods that are used to detect exoplanets, but the physics behind them is beyond the scope of the course.

Astronomy is one of the oldest sciences – there are records of star charts and papers recording the motion of astronomical objects going back to 3000BC. The night sky was observed by Mesopotamians, Egyptians, Greeks, Chinese and Indians. Records exist from these areas dating back thousands of years. In medieval times, records show that a number of Islamic scholars also recorded the movement of the stars and planets.

In European terms, Nicolaus Copernicus was the first to propose a system where all planets orbited the Sun. He used geometrical techniques linked to measurements of the planets' movements to support his ideas. This work was later expanded on by Johannes Kepler, Galileo Galilei and Sir Isaac Newton. Newton described the motion of the stars and planets using his Law of Universal Gravitation.

Our knowledge of the Universe is continually improving through the use of ever more sophisticated telescopes and monitoring devices. We can now monitor accurately the movement of many objects in the Solar System and beyond. Our measurements and observations using telescopes are matched by mathematical descriptions. This allows us to estimate where objects were in the past and predict where they will be in the future.

More recently we have been able to launch objects into space. The first artificial satellite was Sputnik, launched in 1957 by the USSR. We now have launched many satellites that are in space gathering a wide range of data in order for us to improve our understanding of the Universe.

The current position is that the Universe is very large (obviously) and **expanding**. It began from an incredible event almost 14 billion years ago. This event emitted all the matter and energy that exists today. (There is more about this event and its effects in Topic 6.2.)

The current belief that the Universe is expanding is based on the fact that distant galaxies are moving away from us. A more detailed description of the Universe involves understanding some very difficult concepts, such as Einstein's relativity, and adapting our ideas of space and time.

Challenges of space exploration

Space is generally accepted as beginning at a height of 100 km above the surface of the Earth. The energy required to raise an object to this height is vast. Consider a 200 kg satellite. The gain in potential energy needed in order for it to leave the Earth and enter space would be:

$$E_p = mgh$$
$$= 200 \times 9.8 \times 100\,000 = 196\,000\,000 = 200\,000\,000\,J$$

In order for it to remain in space it would have to be in orbit. We would need to provide more energy for it to reach the correct velocity. The energy required makes it difficult and expensive to send large, heavy objects into space.

The Space Shuttle, which had a mass of 100 000 kg, could only achieve heights of about 350 km, which is roughly the orbiting height of the International Space Station (ISS).

At launch, however, the mass of a loaded shuttle and fuel tanks with fuel was around 1 900 000 kg. This enormous load was required to place 100 000 kg in low Earth orbit.

The fuel is *very* dangerous. Liquid hydrogen and liquid oxygen are used in certain rockets and ensuring they are combined correctly is very difficult. On the Challenger shuttle, a flexible ring was used in the connection of one of the fuel tanks. It failed and very hot exhaust gases leaked onto one of the fuel tanks, causing it to explode.

There are other issues to consider in spaceflight.

There is no atmosphere in space and astronauts' spaceships and suits (Figure 6.2) must protect them from this. If they were to be exposed to this vacuum they would lose consciousness and die from lack of oxygen within minutes. The suits and ships provide a breathable mixture of oxygen and other gases. Carbon dioxide in their suits and ships must also be removed so that the air maintains enough oxygen for humans to work in.

The temperature of space is very cold. There can be great changes in the temperature of objects in space depending on whether they are in direct sunlight or shadow. Spaceships and suits must be highly insulated to keep the internal temperature at normal levels.

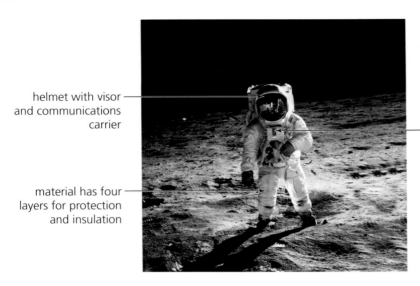

helmet with visor and communications carrier

suit contains pressure control system and oxygen supply

material has four layers for protection and insulation

Figure 6.2 A spacesuit

Satellites and their uses

A satellite, in space terms, is generally defined as an object that orbits a planet. We have one natural satellite, our moon, but also thousands of satellites that humans have launched into space.

A satellite orbits the Earth and does so by having a high velocity. As it travels above the Earth, it is attracted by the Earth's gravity and it falls into an orbit around the Earth.

A very useful orbit is one that travels around the Earth at the same rate as the Earth rotates on its axis. This means that the satellite appears above the same point (on the equator). This allows us to point a receiver to this satellite and establish a link for information to pass. These satellites are geostationary and have to be placed at a height of 36 000 km and at a velocity which means it takes 24 hours to complete one revolution.

When a satellite is in a lower orbit, such as a weather satellite, it has to travel at a greater velocity in order to remain in orbit. (On a clear night you can see these satellites pass by with the naked eye.) Conversely, if a satellite is at a higher orbit it needs to travel at a lower velocity in order to remain there.

Much of life now depends on satellites:
- Global positioning systems (GPS) allow us to use a phone or other device to determine our location to within a few metres.
- Television networks rely heavily on satellites to transfer signals from one area to another, such as live reporting from major events.
- Our weather forecasts are based on data taken from satellite systems which have monitored the area around where we live. We can receive very up to date images of clouds and such which are then shown on our forecasts.
- Satellites with various detectors and telescopes can observe distant objects and allow us to analyse them in order to increase our knowledge of the Universe in which we live. The Hubble space telescope has increased our knowledge of space by a great amount.

These are some examples of some of the uses of satellites and their impact on our lives. There are many more and the use of satellites is constantly increasing.

Re-entry

Bringing a spaceship back to Earth has associated difficulties. As indicated before, a spaceship could be orbiting at heights of 300–400 km and travelling very quickly. The now retired Space Shuttle started re-entry at a height of about 125 km and a velocity of approximately 7000 m s^{-1}. It entered the atmosphere with its nose slightly up and performed a series of rolls to generate drag and friction in order to slow it down. It used the ceramic tiles on its base to withstand the friction from the atmosphere. The heat created by this friction was immense and only ceramic materials could withstand such temperatures. The temperature on the edge of the Shuttle could reach 3000 °C. The base glowed and re-radiated a lot of the energy. The ceramic tiles also acted as insulators to stop the heat conducting through to the Shuttle.

The Space Shuttle used specially made tiles to cope with the energy generated by friction when it was returning to Earth. Most other objects re-entering the Earth's atmosphere will 'burn up'. The high temperatures generated at this time cause the material to heat up and evaporate. Most of the material will evaporate but occasionally small pieces will fall to Earth.

Challenges of space travel

Space travel is *not* like it is portrayed in science-fiction movies! The sheer scale of the distances between stars is very difficult to comprehend. Travel to a nearby planet, such as Mars, would take two or three years and, even then, only when the planets' orbits were in alignment.

If we take the distance between the Sun and Earth as one unit (1 AU or 1 astronomical unit), the most distant object launched by humans, the Voyager 1 satellite, is now about 140 AU away. It was launched in 1977! Our nearest star is approximately 270 000 AU from us.

It requires large amounts of energy in fuel to launch objects into space so that they reach high velocities. It is technically difficult and so very expensive to launch heavy, fuel-laden rockets into space. As a result, we need to develop powerful engines to increase the velocity of objects like satellites while using only a small amount of fuel. Ion drive engines are designed to do this. These engines accelerate ions of xenon between two electric plates. The ions are ejected from the satellite at a very high velocity and this provides a small but constant thrust to the satellite, increasing its speed. Ion drive engines don't produce great acceleration, but, at the current time, they give the greatest amount of acceleration for the smallest amount of mass ejected.

The velocity of satellites can be increased in other ways. We can manoeuvre a satellite close to a planet or large asteroid so that it accelerates towards the planet's surface but misses it. As it heads towards the planet, it accelerates due to that body's gravitational field, increasing its velocity. It now flies past the planet at a greater velocity than before. This is referred as using a gravitational slingshot or catapult. As a result, however, the planet slows down a fraction.

There are other challenges for the scientists who design spaceships and other machines to withstand the space environment and to travel massive distances.

Manoeuvring in a frictionless environment is very different to moving on Earth (Figure 6.3). When we fire a thruster to move a satellite in one direction, it accelerates in the direction of thrust. When we stop firing the thruster, the satellite will not stop! It will carry on at that velocity.

If we fire a thruster at the front of the satellite (for example, pointing to the left) it will not only move the satellite to the left, but it will also cause the satellite to rotate about its centre of mass. When the thruster stops firing, the satellite will continue to move to the left and also rotate.

On Earth, friction forces would cause this motion to slow down and stop. In space, this motion continues until some other force is applied. With no friction, the motion continues indefinitely.

 In order for the spacecraft to manoeuvre in an almost frictionless environment, small rockets (thrusters) are used to change the velocity and direction of the spacecraft.

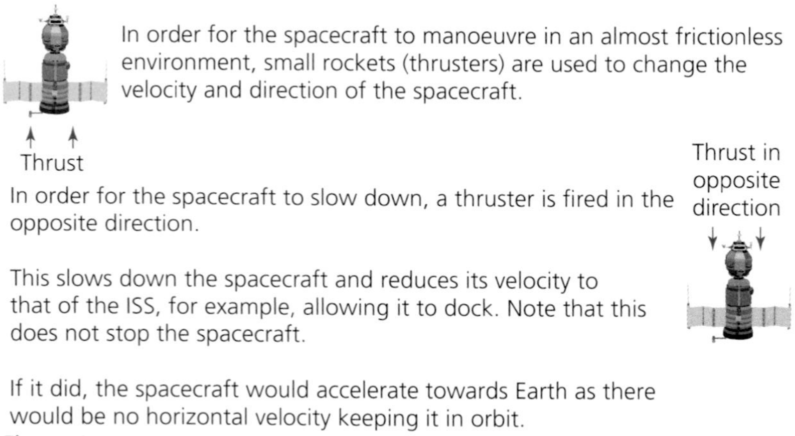

Thrust

In order for the spacecraft to slow down, a thruster is fired in the opposite direction.

Thrust in opposite direction

This slows down the spacecraft and reduces its velocity to that of the ISS, for example, allowing it to dock. Note that this does not stop the spacecraft.

If it did, the spacecraft would accelerate towards Earth as there would be no horizontal velocity keeping it in orbit.

Figure 6.3

Another major issue with long-term space travel is the energy required to power the spacecraft. Batteries are very heavy and cannot store large amounts of energy. We cannot use a simple generator as this would require fuel and oxygen. Solar cells could provide a steady energy supply, but this would diminish as the spacecraft travels further away from our Sun.

Benefits of space exploration

Some people argue that money should be spent on solving the Earth's problems rather than on space exploration. The costs of space exploration are certainly great, but there are also many benefits. In developing the technologies that allow space exploration, many other important developments have been made. Some are obvious, such as weather monitoring satellites and global positioning systems, but there are other benefits too. The following are just a few examples:

- insulation materials and fire-resistant technologies
- infrared aural thermometers
- powdered lubricants
- anti-icing systems for aircraft.

Perhaps the real benefit, however, is the establishment of high-technology programmes that equip people with the skills and ingenuity needed to adapt and develop solutions when faced with new problems.

Questions ?

1   A space vehicle of 18 000 kg returns to Earth. It is travelling at $5000 \, m \, s^{-1}$ when it begins interacting with the atmosphere at a height of 100 km.
   a) Calculate the potential and kinetic energy of the vehicle at that point.
   b) After about 80 minutes the space vehicle lands and is brought to a stop.
      What has most of its energy been converted to?
   c) Describe what properties the base of the space vehicle must have in order for the craft to return safely.
2   List three benefits that satellite technology has brought to your everyday life.
3   The Mars Science Laboratory is the overall title of the project that landed the Mars Curiosity Rover on the surface of Mars in August 2012. The budget for this project is $2·5 billion. Discuss briefly whether you feel it is worth this expense.
4   Research the health issues involved in spending long periods of time in space in a low gravity environment. Choose one topic to focus on and prepare a short paragraph describing the challenges and risks.

6.2
Cosmology

What you should know

For N5 Physics you need to be able to:

★ understand currently held views about the age and origin of the Universe
★ explain the use of the various bands of the electromagnetic spectrum in obtaining information about the Universe
★ identify and explain continuous and line spectra
★ solve simple problems identifying known elements present in stars using line spectra.

Key point

✳ **Cosmology** is the study of the origins and development of the Universe. It is the branch of science that tries to explain the formation of stars, galaxies, space and time.

The most widely accepted theory in cosmology is that the Universe began with a '**Big Bang**'. This term was coined by a British astronomer, Fred Hoyle, who used the term to highlight the difference between the expanding Universe and his own theory of a 'steady state' Universe.

The Big Bang theory is based on observations from telescopes. These observations of distant galaxies all found that they are moving away from us (receding). This can be explained by a single event in which all matter was 'exploded' out from a particular point. Matter that is further away from us is moving away more quickly than we are.

The Big Bang theory predicts that there should be **background microwave radiation** in space, left over from the original event. When this background microwave radiation was first detected in the 1960s, this was seen as strong evidence in favour of the Big Bang theory. While there are small anomalies in the data from various different telescopes, the overwhelming majority of the data supports the Big Bang as the way in which the Universe was created.

The Big Bang is estimated to have occurred just under 14 billion years ago. An incredible amount of energy/matter was released and the 'Universe' grew and expanded. As it expanded the temperature fell as the energy was spread through a greater volume.

Figure 6.4 An artist's impression of the Big Bang

Space itself is still expanding. It is not that the Universe grows while inside some kind of container. The size of the Universe is increasing all the time as it expands. Space is actually being created.

Current observations indicate that the expansion of the Universe is accelerating. It is not known whether this will continue or if it will slow down and become stable. There are many unanswered questions in this area. Scientists make hypotheses, attempt to predict what these hypotheses would mean, and then try to look for evidence to support or refute their theory.

In the 1980s scientists noted that the galaxies were not massive enough to account for the apparent strength of the gravitational forces between them. It was proposed that there was other matter present that could not be observed. This was given the name **dark matter**. The nature of this dark matter and the reasons why it cannot be observed or seen to interact with normal matter is the subject of much study.

Currently the size of the Universe is not known accurately. While we have a reasonable indication of its age, its size could be infinite.

How do we gather information about stars and galaxies?

We gather information about stars and galaxies by observing and detecting the energy they emit. This is done using devices based both here on Earth and in space. We use different sections of the **electromagnetic spectrum** to gather this information.

Stars and galaxies emit energy across the full range of the electromagnetic spectrum, for example as radio waves or X-rays. If we can detect all the energy emitted by a stellar object, we can learn more about the composition and activity of that object. For this we need telescopes that operate across the full range of frequencies of the electromagnetic spectrum. The data gathered from all these different detectors gives a far more detailed picture of the area or object being studied.

Radio astronomy

Radio telescopes detect radiation of wavelengths greater than 10^{-3} m. Radio telescopes are generally large, parabolic dishes, which can be directed towards various different areas of space (Figure 6.5).

Clouds have little effect on radio waves and the telescopes can be used in daytime or at night. They can detect energy from supernovae, pulsars and interstellar gas.

Figure 6.5 A radio telescope

Infrared astronomy

Infrared telescopes detect wavelengths of the order of $1-300 \times 10^{-9}$ m.
Very often optical and infrared telescopes are built into the same housing
and the detectors are adapted to detect different wavelengths. To operate
correctly, infrared telescopes need to be cooled and shielded from warm
objects. They are used to detect energy from 'colder' objects such as dust
and the cores of galaxies. They are affected by atmospheric conditions.

X-ray and gamma ray astronomy

X-ray and gamma ray telescopes detect high energy photons. They are
mainly space-based or high altitude telescopes. Objects such as binary stars,
pulsars, supernovae and neutron stars emit radiation in these wavelengths.

Spectra

Light emitted from a star can be analysed by passing it through a prism or
grating. This separates the light and creates a spectrum (the visible
spectrum). Light from the Sun can be passed through a prism to create a
spectrum, as shown in Figure 6.6.

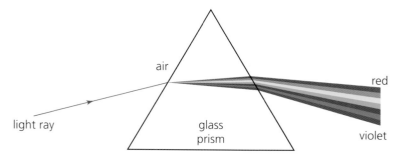

Figure 6.6 Light from the Sun can be separated into the visible spectrum

This is known as a **continuous spectrum**, as there are no gaps and the colours range from red to violet continuously.

Light from distant objects can also be analysed this way. However, it was found that the light from some objects did not split into a continuous spectrum. Instead these objects' lights formed spectra with lines at certain intervals. These are called **line spectra**.

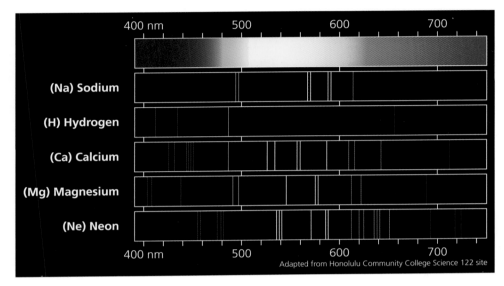

Figure 6.7 Line spectra for sodium, hydrogen, calcium, magnesium and neon

These lines indicate the presence of certain elements (Figure 6.7). They can be used to identify the elements that are present in distant stars.

The light emitted from stars can contain energy from a number of different elements. This means that the spectrum from a particular star can have many lines. Astronomers analyse these spectra and try to 'separate' the lines from all the individual elements to work out the composition of the star.

Observations like these have shown that hydrogen is the most abundant element in the Universe. Helium is the second most abundant. The abundance of these light elements is also strong evidence in favour of the Big Bang.

Questions

1 Why do we use a range of different types of detector when studying stars and galaxies?
2 What do line spectra emitted from a star indicate?
3 The star Proxima Centauri is 4·2 light years from us. Calculate this distance in metres.
4 The star Sirius is 8·6 light years from Earth. If a probe was sent to Sirius, what difficulties would we experience in trying to communicate with this probe?

Solutions to exercises

1
Electricity

1.1 Electrical concepts

Page 5

1 a) i) The electron will move along the field line towards the positive point charge.
   ii) On top of the positive point charge.
   b) i) B
   ii) The neutron does not experience any electrical force in the electric field – it remains at B.
   c) i) The proton will move along the field line towards the negative point charge.
   ii) At the bottom of the negative point charge.

2 a) i) The proton will move down the field line towards the negative point charge on the left.
   ii) On the upper right side of the negative point charge on the left.
   b) i) B
   ii) The neutron does not experience any electrical force in the electric field – it remains at B.
   c) i) The force on the electron is down the field line away from the negative point charge on the right.
   ii) In an electric field the force on a negatively charged object is opposite to the direction of the field.
   iii) The electron is being pushed away from the electric field.

3 For charge to flow in a circuit there must be:
   - a complete path of conductors
   - a source of electrical energy.

4 Conductors are needed to allow charge to flow. Insulators are needed to ensure the safety of users and so that the circuits can be switched on and off.

5 a) electrons
   b) ions (copper: Cu^{2+}; sulphate: SO_4^{2-}; hydrogen: H^+; and hydroxide: OH^-)
   c) electrons and holes

6 Water molecules are continuously breaking down into hydrogen ions, H^+, and hydroxide ions, OH^-, and also recombining to form water molecules. The pure water always contains hydrogen and hydroxide ions and these are the charge carriers.

7 $Q = 600\,C$ $Q = It$
  $t = 0.15\,s$ $\Rightarrow 600 = I \times 0.15$
                      $\Rightarrow I = 4000\,A$

8 The p.d. between the terminals of the supply $= 15\,V$. (*When* $Q = 1\,C$, $V = W$.)

9 $Q = 40\,mC = 4.0 \times 10^{-2}\,C$
  $V = 1.8\,kV = 1.8 \times 10^3\,V$ (*Careful with the units!*)
  $E_w = ?$ $E_w = QV$
                                $= 4.0 \times 10^{-2} \times 1.8 \times 10^3$
                                $= 72\,J$

10 a) $Q = 0.30\,C$ $E_w = QV$
   $E_w = 1.2\,J$ $\Rightarrow 1.2 = 0.30 \times V$
   $V = ?$ $\Rightarrow V = 4.0\,V$
   b) E_w moving $0.30\,C$ from A to D to C $= 1.2\,J$. The work done in moving charge in an electric field depends only on the starting and finishing points.

11 a) (*Kinetic energy gained by electrons = work done.*)
   $Q = 1.6 \times 10^{-19}\,C$ $E_k = E_w = QV$
   $V = 2.30\,kV$ $\Rightarrow E_k = 1.6 \times 10^{-19} \times 2.3 \times 10^3$
       $= 2.3 \times 10^3\,V$ $= 3.68 \times 10^{-16}\,J$
                                   $= 3.7 \times 10^{-16}\,J$

b)
$$m_e = 9{\cdot}11 \times 10^{-31}\,\text{kg}$$
$$E_k = \frac{1}{2}mv^2$$
$$\Rightarrow 3{\cdot}68 \times 10^{-16} = \frac{1}{2} \times 9{\cdot}11 \times 10^{-31} \times v^2$$
$$\Rightarrow v = 2{\cdot}842 \times 10^7 = 2{\cdot}8 \times 10^7\,\text{m s}^{-1}$$

c) Assumption: the electrons are initially at rest.

12

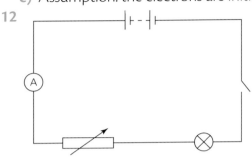

(The ammeter may be connected in series anywhere in the circuit.)

13

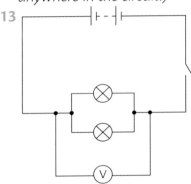

(The voltmeter may be shown anywhere provided that it is in parallel with the lamps; the voltage across both lamps is the same.)

14

Component	Symbol	What it does
cell	⊢⊢	Source of electrical energy
ammeter	—Ⓐ—	Measures current in electrical circuits
resistor	⊢▭⊢	Opposes flow of charge in a circuit
supply	—o o—	Source of electrical energy
switch	⟋ —	Turns current on and off
fuse	⊢▭⊢	Stops current when it gets too large
lamp	—⊗—	Source of light
voltmeter	—Ⓥ—	Measures p.d. across components
variable resistor	⊢▱⊢	Increases or decreases current
battery	⊢⊢-⊢⊢	Source of electrical energy

1.2 Electrical circuits

Page 12

1 a) *(I in 30 Ω resistor = I in 18 Ω resistor = 0·25 A.)*

$V = ?$	$V = IR$
$I = 0{\cdot}25\,\text{A}$	$\Rightarrow V = 0{\cdot}25 \times 30$
$R = 30\,\Omega$	$= 7{\cdot}5\,\text{V}$

b)

$R_{circuit} = ?$	$V = IR$
$I = 0{\cdot}25\,\text{A}$	$\Rightarrow 12 = 0{\cdot}25 \times R$
$V = 12\,\text{V}$	$\Rightarrow R_{circuit} = 48\,\Omega$

(Note: circuit resistance = sum of individual resistances = 18 + 30 = 48 ohms.)

2 a)

$I = ?$	$V = IR$
$V = 6{\cdot}0\,\text{V}$	$\Rightarrow 6{\cdot}0 = I \times 60$
$R = 60\,\Omega$	$\Rightarrow I = 0{\cdot}10\,\text{A}$

b) Parallel circuit so current in R_2 = current in battery – current in 60 Ω resistor.
$\Rightarrow$ current in $R_2 = 0{\cdot}30 - 0{\cdot}10 = 0{\cdot}20\,\text{A}$

$R_2 = ?$	$V = IR$
$I = 0{\cdot}20\,\text{A}$	$\Rightarrow 6{\cdot}0 = 0{\cdot}20 \times R$
$V = 6{\cdot}0\,\text{V}$	$\Rightarrow R_2 = 30\,\Omega$

3 a) The current in the battery = 2·0 A. The battery and the 2·0 Ω resistor are in series.

b) *(First calculate the p.d. across the 2·0 Ω resistor.)*

$V = ?$	$V = IR$
$I = 2{\cdot}0\,\text{A}$	$\Rightarrow V = 2{\cdot}0 \times 2{\cdot}0$
$R = 2{\cdot}0\,\Omega$	$= 4{\cdot}0\,\text{V}$

(The path through the 2·0 Ω and the 16 Ω resistor makes a complete path around the circuit.)

$\Rightarrow$ p.d. across 16 Ω resistor = battery voltage – p.d. across 2·0 Ω resistor
$$= 20 - 4{\cdot}0 = 16\,\text{V}$$

c) *(First calculate the current in the 16 Ω resistor.)*

$I = ?$	$V = IR$
$V = 16\,\text{V}$	$\Rightarrow 16 = I \times 16$
$R = 16\,\Omega$	$\Rightarrow I = 1{\cdot}0\,\text{A}$

Now current in R_3 = current in 2·0 Ω resistor – current in 16 Ω resistor
$$= 2{\cdot}0 - 1{\cdot}0 = 1{\cdot}0\,\text{A}$$

$R_3 = ?$	$V = IR$
$I = 1{\cdot}0\,\text{A}$	$\Rightarrow 16 = 1{\cdot}0 \times R_3$
$V = 16\,\text{V}$	$\Rightarrow R_3 = 16\,\Omega$

1.3 Resistors

Page 16

1 a) Your graph should look like this:

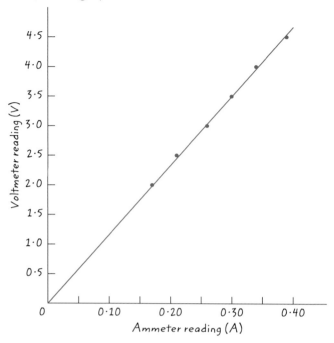

Use the origin and any point **on the line** — for example (0·345, 4·0):
$$R_{resistor} = \text{gradient of graph} = \frac{4\cdot0}{0\cdot345} = 11\cdot594 = 11\cdot6\,\Omega$$

b) When the switch is open the reading on both meters is zero.

2 a) $R_1 = 10\,\Omega$ $R_T = ?$
$R_2 = 10\,\Omega$ $R_T = R_1 + R_2 \ldots + R_{10}$
$R_3 = 10\,\Omega$, etc. $\Rightarrow R_T = 10 + 10 \ldots + 10$
$\qquad\qquad\qquad = 100\,\Omega$

b) $R_1 = 10\,\Omega$ $\dfrac{1}{R_T} = \dfrac{1}{R_1} + \dfrac{1}{R_2} \ldots + \dfrac{1}{R_{10}}$
$R_2 = 10\,\Omega$
$R_3 = 10\,\Omega$, etc. $\Rightarrow \dfrac{1}{R_T} = \dfrac{1}{10} + \dfrac{1}{10} \ldots + \dfrac{1}{10}$
$R_T = ?$ $\qquad\qquad = 1\cdot0$
$\qquad\qquad\qquad \Rightarrow R_T = 1\cdot0\,\Omega$

3 a) (First find the resistance of the parallel branches.)
$R_1 = 8\cdot0\,\Omega$ $\dfrac{1}{R_T} = \dfrac{1}{R_1} + \dfrac{1}{R_2}$
$R_2 = 12\cdot0\,\Omega$
$R_T = ?$ $\Rightarrow \dfrac{1}{R_T} = \dfrac{1}{8\cdot0} + \dfrac{1}{12\cdot0}$
$\qquad\qquad\qquad = 0\cdot2083$
$\qquad\qquad\qquad \Rightarrow R_T = 4\cdot8\,\Omega$

(Always remember to invert after the addition!)

(Now add to the other resistance.)

$R_1 = 4\cdot8\,\Omega$ $R_T = R_1 + R_2$
$R_2 = 5\cdot2\,\Omega$ $\Rightarrow R_T = 4\cdot8 + 5\cdot2$
$\qquad\qquad\qquad = 10\,\Omega$

b) (First find the current across the 5·2 Ω resistor.)
$I_{battery} = ?$ $V = IR$
$\quad V = 12\,V$ $\Rightarrow 12 = I \times 10$
$\quad R = 10\,\Omega$ $\quad I = 1\cdot2\,A$

(Now find the p.d. across the 5·2 Ω resistor.)
$V_{5\cdot2\,\Omega} \text{ resistor} = ?$ $V = IR$
$I = 1\cdot2\,A$ $\Rightarrow V = 1\cdot2 \times 5\cdot2$
$R = 5\cdot2\,\Omega$ $\qquad = 6\cdot24\,V$

(Now subtract this from the battery voltage.)
p.d. across 8·0 Ω resistor $= 12 - 6\cdot24 = 5\cdot76$
$\qquad\qquad\qquad = 5\cdot8\,V$

4 a) (First find the resistance of the two resistors in series.)
$R_1 = 8\cdot0\,\Omega$ $R_T = R_1 + R_2$
$R_2 = 10\cdot0\,\Omega$ $\Rightarrow R_T = 8\cdot0 + 10\cdot0 = 18\,\Omega$
$R_T = ?$

(Now find the resistance of the parallel branches.)
$R_1 = 18\,\Omega$ $\dfrac{1}{R_T} = \dfrac{1}{R_1} + \dfrac{1}{R_2}$
$R_2 = 9\cdot0\,\Omega$
$R_T = ?$ $\Rightarrow \dfrac{1}{R_T} = \dfrac{1}{18} + \dfrac{1}{9\cdot0}$
$\qquad\qquad\qquad = 0\cdot166$
$\qquad\qquad\qquad \Rightarrow R_T = 6\cdot0\,\Omega$

(Parallel resistors so invert after adding!)

b) (First calculate the current in the branch containing the 8·0 Ω resistor.)
$I = ?$ $V = IR$
$V = 24\,V$ $\Rightarrow 24 = I \times 18$
$R = 18\,\Omega$ $\Rightarrow I = 1\cdot33\,A$

(Now calculate the p.d. across the 8·0 Ω resistor.)
$V = ?$ $V = IR$
$I = 1\cdot33\,A$ $\Rightarrow V = 1\cdot33 \times 8\cdot0$
$R = 8\cdot0\,\Omega$ $\qquad = 10\cdot67 = 11\,V$

5 a) (Voltage between points A and B = p.d. across the 100 Ω resistor.)
$R_1 = 80\,\Omega$
$R_2 = 100\,\Omega$ $V_2 = \left(\dfrac{R_2}{R_1 + R_2}\right)V_s$
$V_s = 9\cdot0\,V$
$V_2 = ?$ $\Rightarrow V_2 = \left(\dfrac{100}{80 + 100}\right) \times 9\cdot0$
$\qquad\qquad\qquad \Rightarrow V_2 = 5\cdot0\,V$

b) p.d. across the $80\,\Omega$ resistor $= 4\cdot0\,V$. It equals supply voltage minus $5\cdot0\,V$.

6 a) (First calculate the resistance of each branch.)

$R_1 = 30\,\Omega$ $\qquad$ $R_{12} = R_1 + R_2 = 30 + 60 = 90\,\Omega$

$R_2 = 60\,\Omega$

$R_3 = 20\,\Omega$ $\qquad$ $R_{34} = R_3 + R_4 = 20 + 40 = 60\,\Omega$

$R_4 = 40\,\Omega$

(Now calculate the total resistance of the parallel branches.)

$$\frac{1}{R_T} = \frac{1}{R_{12}} + \frac{1}{R_{34}} = \frac{1}{90} + \frac{1}{60} = 0\cdot277$$

$$\Rightarrow R_T = 36\,\Omega$$

b) $V = 6\cdot0\,V$ $\qquad\qquad V = IR$

$R_T = 36\,\Omega$ $\qquad\quad \Rightarrow 6\cdot0 = I \times 36$

$\qquad\qquad\qquad\qquad \Rightarrow \quad I = 0\cdot166$

$\qquad\qquad\qquad\qquad\qquad = 0\cdot17\,A$

c) (Current in $60\,\Omega$ resistor = current in its branch.)

$V = 6\cdot0\,V$ $\qquad\qquad V = IR$

$R_{12} = 90\,\Omega$ $\qquad \Rightarrow 6\cdot0 = I \times 90$

$\qquad\qquad\qquad\qquad \Rightarrow \quad I = 0\cdot0666$

$\qquad\qquad\qquad\qquad\qquad = 0\cdot067\,A$

d) (Current in $20\,\Omega$ resistor = current in its branch.)

$V = 6\cdot0\,V$ $\qquad\qquad V = IR$

$R_{34} = 60\,\Omega$ $\qquad \Rightarrow 6\cdot0 = I \times 60$

$\qquad\qquad\qquad\qquad \Rightarrow \quad I = 0\cdot1\,A$

e) (Left sides of $20\,\Omega$ and $30\,\Omega$ resistors are connected to the same point so they are at the same voltage – so calculate the p.d.s across the $20\,\Omega$ and $30\,\Omega$ resistors.)

$I = 0\cdot0666\,A$ $\qquad\qquad V = IR$

$R_1 = 30\,\Omega$ $\qquad\quad \Rightarrow V = 0\cdot0666 \times 30$

$\qquad\qquad\qquad\qquad\qquad = 1\cdot998 = 2\cdot0\,V$

$I = 0\cdot1\,A$ $\qquad\qquad\quad V = IR$

$R_1 = 20\,\Omega$ $\qquad\quad \Rightarrow V = 0\cdot1 \times 20$

$\qquad\qquad\qquad\qquad\qquad = 2\cdot0\,V$

(Right sides of the $20\,\Omega$ and $30\,\Omega$ resistors are at the same voltage.)

$\Rightarrow$ p.d. between X and Y $= 0\,V$

1.4 Electrical energy
Page 21

1 In an electric fire the energy change takes place in the resistance wire (element).

2 The energy change in a filament lamp is electrical energy $\rightarrow$ light + heat.

3 a) $P = IV$ and $V = IR$

Substitute for $V \Rightarrow P = I \times IR = I^2R$

b) $P = IV$ and $V = IR \Rightarrow I = \dfrac{V}{R}$

Substitute for $I \Rightarrow P = \dfrac{V}{R} \times V = \dfrac{V^2}{R}$

4 a) $P = 3\cdot0\,kW = 3000\,W$ $\qquad P = IV$

$V = 230\,V$ $\qquad\qquad \Rightarrow 3000 = I \times 230$

$I = ?$ $\qquad\qquad\qquad \Rightarrow \quad I = 13\cdot04$

$\qquad\qquad\qquad\qquad\qquad\qquad = 13\,A$

b) $P = 3000\,W$ $\qquad\qquad E = Pt$

$t = 5$ minutes $= 300\,s$ $\quad \Rightarrow E = 3000 \times 300$

$E = ?$ $\qquad\qquad\qquad\qquad = 9\cdot0 \times 10^5\,J$

5 Four identical elements so power used by each element $= \dfrac{1840}{4} = 460\,W$

$P = 460\,W$ $\qquad\qquad\qquad P = \dfrac{V^2}{R}$

$V = 230\,V$ $\qquad\qquad\qquad \Rightarrow 460 = \dfrac{230^2}{R}$

$R = ?$

$\qquad\qquad\qquad\qquad\qquad \Rightarrow \quad R = 115\,\Omega$

6 a) Appropriate fuse is $13\,A$

b) Appropriate fuse is $3\,A$

c) Appropriate fuse is $13\,A$

d) Appropriate fuse is $3\,A$

7 a) The $13\,A$ fuse in the extension lead should 'blow', cutting off the current.

b) $P = 1300 + 2500 = 3800\,W$ $\qquad P = IV$

$V = 230\,V$ $\qquad\qquad\qquad \Rightarrow 3800 = I \times 230$

$I = ?$ $\qquad\qquad\qquad\qquad \Rightarrow \quad I = 16\cdot5\,A$

This current is too great for the 13 amp fuse protecting the extension lead.

1.5 Semiconductor diodes

Page 25

1 a) Gallium (atomic number 31), because gallium is next to germanium (atomic number 32) in the Periodic Table. Its atoms are about the same size as germanium atoms and have three electrons in the outer shell.

   b) Arsenic (atomic number 33), because arsenic is next to germanium (atomic number 32) in the Periodic Table. Its atoms are about the same size as germanium atoms and have five electrons in the outer shell.

2 The p-type part of the depletion layer is negatively charged and the n-type part of the depletion layer is positively charged.

3 Sample answers: cheaper, more reliable, lighter, smaller.

1.6 Capacitors

Page 29

1 a) The ammeter reading starts high and gradually falls to zero. The voltmeter reading starts high and gradually falls to zero.

   b)

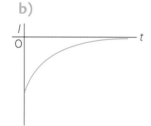

2 a) The potential difference between the plates of a capacitor is directly proportional to the excess charge on each plate.

   b) The plates of a capacitor contain many millions of charges – that is, protons and electrons. In neutral plates there are equal numbers of electrons and protons. A negatively charged plate has more electrons than protons and a positively charged plate has more protons than electrons.

1.7 Electrical supplies for electronic circuits

Page 33

1 In a.c. the charge changes direction. In d.c. the charge moves in one direction.

2
$n_s = 200$ turns

$n_p = ?$

$V_s = 6.3\,V$

$V_p = 230\,V$

$V_s = \dfrac{n_s}{n_p}V_p$

$\Rightarrow 6.3 = \dfrac{200}{n_p} \times 230$

$\Rightarrow n_p = 7302 = 7300$ turns

3 a) B→O→P→R→N→M→A

   b) A→M→P→R→N→O→B

4 Half wave rectifiers are very inefficient – at least half of the input energy is wasted.

1.8 Input and output devices

Page 37

1 a) Thermistor

   b) Thermistor or thermocouple

   c) Thermocouple

   d) LDR

2 Sample answers:

   a) Motors are used in model cars, vacuum cleaners, CD players, DVD players ... etc.

   b) LEDs are used as indicator lights to show when electrical equipment is on, in traffic lights ... etc.

   c) Buzzers are used to attract attention as doorbells, in alarm clocks, in quiz shows ... etc.

   d) Loudspeakers are used in radios, CD players, televisions, public address systems ... etc.

3 a) Energy change in an electric motor is electrical energy → kinetic energy.

   b) Energy change in a thermistor is heat → electrical energy.

   c) Energy change in a buzzer is electrical energy → sound.

   d) Energy change in an LDR is light → electrical energy.

4   All input devices change other forms of energy to electrical energy.

5   a) Component X is a thermistor.

b) $R_1 = 5.0\,k\Omega$
$R_2 = 2.5\,k\Omega$
$V_s = 6.0\,V$
$V_2 = ?$

$$V_2 = \left(\frac{R_2}{R_1 + R_2}\right)V_s$$

$$\Rightarrow V_2 = \left(\frac{2.5}{5.0 + 2.5}\right) \times 6.0$$

$$= 2.0\,V$$

(In the question above all resistances were quoted in kΩ — it was not necessary to change these to Ω because the term in brackets has kΩ divided by kΩ!)

c) This type of circuit could be used to switch off an oven when it reaches a particular temperature.

1.9 Switches

Page 40

1   a)

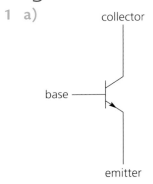

base — collector, emitter

b)

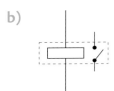

2   a) Component X is an n-channel enhancement MOSFET.

b) Consider the potential divider circuit. During the hours of darkness the p.d. across the LDR is exactly half of the supply voltage. Hence resistance of R = resistance of LDR = 5 kΩ.

c) As the intensity of light incident on the LDR increases, the resistance of the LDR falls. This causes the p.d. across the LDR to decrease.
The MOSFET switches off when the p.d. across the LDR falls below a set value (or 2 V). This switches off the relay.

2

Properties of matter

2.1 Specific heat capacity and latent heat

Page 43

1   $E_h = 128 \times 15 \times 25 = 48\,000 = 48\,000\,J$

2   $E_h = 4180 \times 0.15 \times 35 = 21\,945 = 22\,000\,J$

3   $c = \dfrac{32\,000}{2 \times 8} = 2000\,J/kg/C$

4   $m = \dfrac{16\,000}{480 \times 32} = 1.04\,kg$

5   $E_h = 4180 \times 2.0 \times 75 = 627\,000 = 630\,000\,J$

Page 44

1   $E_h = ml = 0.45 \times 3.34 \times 10^5 = 1.5 \times 10^5\,J$

2   $l = \dfrac{E_h}{m} = \dfrac{240\,000}{1.3} = 184\,615 = 1.8 \times 10^5\,J/kg$

2.2 Gas laws and the kinetic model

Page 46

1   a) $W = mg = 1500 \times 9.8 = 14\,700\,N$

$$P = \frac{F}{A} \Rightarrow P = \frac{14\,7000}{0.9} = 1.633 \times 10^4\,Pa$$

b) $P = \frac{F}{A} = \frac{147000}{0.4 \times 0.4}$

$= 91875$

$= 9.2 \times 10^4 \, Pa$

2 $W = mg = 80 \times 9.8 = 784 \, N$

$P = \frac{F}{A} = \frac{784}{0.025} = 3.136 \times 10^4 \, Pa = 3.1 \times 10^4 \, Pa$

3 $P = \frac{F}{A}$

$\Rightarrow 12000 = \frac{F}{0.15}$

$F = 12000 \times 0.15$

$= 1800 \, N$

4 Chair legs have small surface areas. If a person tilts the chair back the base of the leg will not be flat on the floor. Only a section of it will be on the floor. This means a small surface area will be in contact with the floor and this will create a great pressure on the floor, possibly causing a dent or some damage.

5 Grass can be a slippery surface so the blades/studs have small surface areas in contact with the ground. This causes a high pressure which can bite into the turf and give better grip.

6 Heavy loads need to have their weight distributed over a large area to reduce pressure. Large transport trucks have many wide tyres (about 18) which distribute the load over a large area, reducing the pressure.

Page 50

1 $\frac{V_1}{T_1} = \frac{35}{294} = 0.119 \qquad T_2 = \frac{V_2}{0.119}$

$\frac{V_2}{T_2} = 0.119 \qquad\qquad = \frac{48}{0.119} = 403 \, k = 130\,°C$

2 $\frac{V_1}{T_1} = \frac{120}{307} = 0.391 \qquad T_2 = \frac{V_2}{0.391}$

$\frac{V_2}{T_2} = 0.391 \qquad\qquad = \frac{92}{0.391} = 235 \, k = -37.7\,°C$

$= -38\,°C$

$\Rightarrow 240000 = 1.3 \times L$

$\Rightarrow \qquad L = \frac{240000}{1.3}$

$= 1.85 \times 10^5 \, Jkg^{-1}$

$= 1.8 \times 10^5 \, Jkg^{-1}$

Page 51

1

Celsius	−273	−200	−73	0	27	150	327	500
Kelvin	0	73	200	273	300	423	600	773

2 $4.9 \times 10^5 \, Pa$

3 $5.0 \times 10^4 \, Pa$

4

Volume (cm³)	210	240	260	284	310	360	500
Temperature (°C)	−34	0	23	50	80	137	296

5 $2.57 m^3$

6 The temperature of the air in the tyres has increased. This increases the average speed of the molecules, which increases the number of collisions within the tyre walls, resulting in an increase in pressure.

3
Waves

3.1 Properties of waves

Page 54

1 30 Hz has a wavelength of 11.3 m. 17 500 Hz has a wavelength of 0.02 m.

2 $\frac{340}{25000} = 0.0136 = 0.014$

3 $3 \times 10^{-4} \, m$

4 $5 \times 10^9 \, Hz$

5 1800 MHz is 0.17 m and 1900 MHz is 0.16 m.

3.2 Electromagnetic spectrum

Page 58

1 Answer depends on wavelength chosen.

2 a) Gamma rays are used in the treatment of tumours. Rays are concentrated on one spot and supply a lethal dose to the tumour.

   b) Thermal imaging is used by fire safety crews to try and find people trapped in collapsed buildings, for example.

c) UV radiation on our skin helps produce vitamin D which is a key component of good health. However, too much UV can lead to sunburn, swelling, wrinkled skin and cancer.

3 The frequency range of X-rays is taken to be approximately 3×10^{16} Hz to 3×10^{19} Hz. Lower frequency X-rays are used in human X-rays but higher frequency radiation is used in airport security scanners where no living organism is affected by the rays.

4 Debate here could include: the advertising of high-power tubes to tan more quickly; personal choice if people want to look tanned and know the risks; existence of tanning salons may encourage a self-harmful behaviour.

5 The risk from X-rays is much less than that of an undiagnosed broken bone or damaged organ.

6 Almost all reports have said there is no direct link between mobile phone use and any cancer. Some have suggested that in all things it is better to reduce exposure to radiations of any kind as a general rule.

3.3 Light
Page 61
1
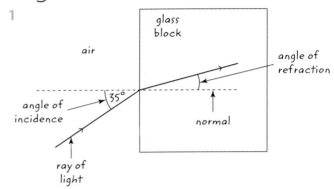

2 The light from an immersed object has to cross the water/air boundary. This causes the ray of light to change direction (bend).

3 Any three from: telescope; microscope; magnifying glass; camera; data projector; binoculars.

4
Radiation

4.1 Nuclear radiation
Page 65

1 a) 80 kg man would absorb 2.5×10^{-4} Gy. 0.5 kg hamster would absorb 0.04 Gy.
   b) For the man: 5×10^{-3} Sv. For the hamster: 0.8 Sv.
   c) It has a much smaller mass.

2 An alpha particle is composed of 2 protons and 2 neutrons. Its charge is positive. It is not very penetrating and can be stopped by a few sheets of paper or about 30–40 cm of air. If ingested it can be harmful.

3 Radiation in the upper atmosphere is higher than at ground level due to cosmic radiation.

4 Radon, granite.

5 There is a large amount of granite in the Aberdeen area and this is slightly radioactive.

Page 72
1 Depends on student's response.
2 Depends on student's response.
3 a) 56 g
   b) 4 g

4
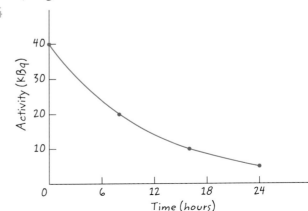

5 Uranium
6 Deuterium and tritium
7 Nuclear power produces little or no carbon emissions compared to fossil fuel power stations.
8 Nuclear waste may remain radioactive for very long periods. The storage and maintenance of the containers will be an issue for hundreds, possibly thousands, of years. Will the

radioactivity from the waste affect the tunnels or mines it is stored in?
Will the containers be able to withstand the radioactivity for such a long period of time?
9 a) 1·5 days
 b) 62·5 kBq
10 2 hours
11 22 920 years

5
Dynamics

5.1 Describing movement

Page 78

1 a) Distance, speed and time are scalars.
 b) Velocity, displacement and acceleration are vectors.
2 a) Force and weight have directions.
 b) Force and weight are vector quantities as they have both size and direction.
3 The boy is not correct. When an object moves in a straight line **without changing direction**, the size of the distance and displacement are always the same. If the object changes direction, it could still be moving in a straight line after the change but the distance and displacement would be different.
4 a) $v = ?$ $\qquad d = vt$
 $d = 4.0\,\text{m}$ $\qquad \Rightarrow 4.0 = v \times 0.080$
 $t = 0.080\,\text{s}$ $\qquad \Rightarrow v = 50\,\text{m}\,\text{s}^{-1}$
 b) This is an instantaneous speed — it is calculated over a very short period of time.
5 a) $d = 2 \times 400 = 800\,\text{m}$ $\qquad d = \bar{v}t$
 $t = (2 \times 60) + 8 = 128\,\text{s}$ $\qquad \Rightarrow 800 = \bar{v} \times 128$
 $\qquad\qquad\qquad\qquad\qquad \Rightarrow \bar{v} = 6.25\,\text{m}\,\text{s}^{-1}$
(Final answer has 3 significant figures as data in the question have 3 significant figures.)
 b) Her average velocity = her final displacement divided by her time.
 The girl started and ended the race at the finish line so her final displacement = 0 m.
 $\Rightarrow$ her average velocity for the race = 0 m s^{-1}.

6 a) $\bar{v} = ?$ $\qquad\qquad\qquad d = \bar{v}t$
 $d = 360\,\text{m}$ $\qquad\qquad \Rightarrow 360 = \bar{v} \times 180$
 $t = 180\,\text{s}$ $\qquad\qquad \Rightarrow \bar{v} = 2.0\,\text{m}\,\text{s}^{-1}$
 b) $\bar{v} = ?$ $\qquad\qquad\qquad d = \bar{v}t$
 $d = 360\,\text{m}$ $\qquad\qquad \Rightarrow 360 = \bar{v} \times 120$
 $t = 120\,\text{s}$ $\qquad\qquad \Rightarrow \bar{v} = 3.0\,\text{m}\,\text{s}^{-1}$
7 $d = ?$ $\qquad\qquad\qquad\qquad d = \bar{v}t$
 $\bar{v} = 25\,\text{m}\,\text{s}^{-1}$ $\qquad\qquad = 25 \times 1800$
 $t = 30\,\text{min} = 1800\,\text{s}$ $\qquad = 45\,000\,\text{m}$
 $\qquad\qquad\qquad\qquad\qquad\qquad (= 45\,\text{km})$
(You can leave the final answer in metres — convert to km only if the question asks for the answer in km.)
8 $t = ?$ $\qquad\qquad\qquad\qquad d = \bar{v}t$
 $\bar{v} = 3.8\,\text{m}\,\text{s}^{-1}$ $\qquad\qquad \Rightarrow 11\,400 = 3.8 \times t$
 $d = 11.4\,\text{km} = 11\,400\,\text{m}$ $\Rightarrow \qquad t = 3000\,\text{s}\,(= 50\,\text{min})$
(You can leave the final answer in seconds — convert to minutes only if the question asks for the answer in minutes.)
9 Calculate the total time to complete 800 m at 5 m s^{-1}:
 $t = ?$ $\qquad\qquad\qquad\qquad d = \bar{v}t$
 $\bar{v} = 5.0\,\text{m}\,\text{s}^{-1}$ $\qquad\qquad \Rightarrow 800 = 5.0 \times t$
 $d = 800\,\text{m}$ $\qquad\qquad\qquad \Rightarrow t = 160\,\text{s}$
The girl has taken 160 s to run the first 400 m. She still has 400 m to run and no time left, so she cannot reach an average speed of 5·0 m s^{-1} for the whole distance.
10 a) The students:
 • use a metre stick to measure the diameter of the ball — at least five measurements should be made and an average calculated

- set up a light gate, connected to a computer, level with the centre of the stationary ball (the light gate should be in front of and close to the ball – when the boy kicks the ball he has to make sure that it goes through the light gate)
- use the light gate and computer to measure the time for the ball to pass through the light gate – at least five measurements should be taken and an average time calculated
- find the instantaneous speed of the ball by dividing the average diameter of the ball by the average time for the ball to pass through the light gate.

b) The students:
- use a metre stick to measure the distance between the ball and the target – at least five measurements should be made and an average distance calculated
- use a stopwatch to measure the time taken
- start the watch as the ball is kicked and stop it when the ball hits the target – at least five measurements should be taken and an average time calculated
- find the average speed of the ball by dividing the average distance by the average time.

c) The instantaneous speed is likely to be greater than the average speed. Air resistance may cause the ball to slow down after it has been kicked.

11 $u = 0\,\text{m s}^{-1}$
$v = 25\,\text{m s}^{-1}$
$t = 10\,\text{s}$
$a = ?$

$a = \dfrac{v - u}{t}$

$\Rightarrow a = \dfrac{25 - 0}{10}$

$= 2.5\,\text{m s}^{-2}$

12 $u = 10\,\text{m s}^{-1}$
$v = 0\,\text{m s}^{-1}$
$t = 4.0\,\text{s}$
$a = ?$

$a = \dfrac{v - u}{t}$

$\Rightarrow a = \dfrac{0 - 10}{4}$

$= -2.5\,\text{m s}^{-2}$

13 $u = 3.4\,\text{m s}^{-1}$
$v = ?$
$t = 3.0\,\text{s}$
$a = 0.20\,\text{m s}^{-2}$

$a = \dfrac{v - u}{t}$

$\Rightarrow 0.20 = \dfrac{v - 3.4}{3}$

$\Rightarrow \quad v = 4.0\,\text{m s}^{-1}$

14 $a = 25\,\text{m s}^{-2}$
$t = 2 \times 60 = 120\,\text{s}$

$a = \dfrac{\Delta v}{t}$

$\Rightarrow 25 = \dfrac{\Delta v}{120}$

$\Rightarrow \Delta v = 3000\,\text{m s}^{-1}$

5.2 Adding vectors

Page 81

1 Velocity of boat through water $= 5.0\,\text{m s}^{-1}$
Velocity of water $= 3.5\,\text{m s}^{-1}$

$v^2 = 5.0^2 + 3.5^2$

$= 25 + 12.25 = 37.25$

$\Rightarrow$ size of resultant velocity $= 6.1\,\text{m s}^{-1}$
Opposite $= 5.0\,\text{m s}^{-1}$
Adjacent $= 3.5\,\text{m s}^{-1}$

$\tan\theta = \dfrac{\text{opposite}}{\text{adjacent}}$

$= \dfrac{5.0}{3.5} = 1.43$

$\Rightarrow \theta = 55°$

$\Rightarrow$ resultant velocity of boat $= 6.1\,\text{m s}^{-1}$ at $55°$ to the bank

2 a) Distance $d = 300 + 450 + 450 + 200 = 1400\,\text{m}$ (= $1.4\,\text{km}$)

b) Resultant $d = 300 - 450 - 450 + 200 = -400\,\text{m}$ ($400\,\text{m}$ east)

5.3 Velocity–time graphs

Page 84

1 a) The initial velocity is not zero – the graph does not start at $v = 0$.

b) i) The velocity of the object is increasing between points AB, EF and FG.

ii) The velocity of the object is decreasing between points CD and GH.

iii) The velocity of the object is constant between points BC and DE.

c) The final velocity of the object is $0\,\text{m s}^{-1}$.

2 a) i) The velocity of the object is positive between points CH.

ii) The velocity is negative between points OC and HK.

b) The velocity of the object is zero at points O, C, H and K.

c) The acceleration of the object is:
 i) positive between points BD and JK
 ii) negative between points OA, EF and GI
 iii) zero between points AB, DE, FG and IJ.

3

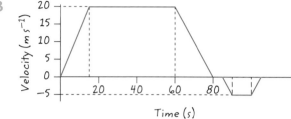

4 a) d = area under speed-time graph
 $= \frac{1}{2} \times 5 \times 10$ (area of triangle) = 25 m

 b) d = area under speed-time graph
 $= 25 + (10 \times 10)$ (add area of rectangle) = 125 m

 c) i) d = area under speed-time graph
 $= (\frac{1}{2} \times 5 \times 2) + (5 \times 8)$ (area of triangle +
 area of rectangle)
 $= 5 + 40 = 45$ m

 ii) The cyclist could be slowing down because
 he is approaching a bend in the track or
 due to air resistance, getting tired, etc.

5 a) The ball is dropped at $t = 0$ s. It falls and
 bounces at $t = 1$ s. The ball rises for 0·8 s,
 stops at its highest point momentarily and
 begins to fall again. At $t = 2$ s the ball is falling.

 b) Initial height of ball = displacement after 1 s
 = area under velocity–time graph
 $= \frac{1}{2} \times 1 \times 9·8 = 4·9$ m

 c) Initial acceleration of ball = a during first
 second
 = gradient of velocity–time graph
 $= \frac{9·8 - 0}{1} = 9·8$ m s^{-2}

 d) Height of first bounce = displacement
 between $t = 1$ s and $t = 1·8$ s
 = area under velocity–time graph
 $= \frac{1}{2} \times -8 \times 0·8 = -3·2$ m
 Height of first bounce = 3·2 m
 (*Normally the negative sign must be included
 in the statement of the final answer; in this
 case, the wording of the question means that
 the negative sign may be omitted. If you are
 in any doubt include the negative sign.*)

 e) Down is the positive direction. The initial
 movement of the ball is downwards and the
 initial part of the graph is positive.

6 a) $a = ?$
 $v = 30$ m s^{-1}
 $u = 0$ m s^{-1}
 $t = 7·9$ s
 $$a_{carA} = \frac{v - u}{t}$$
 $$= \frac{30 - 0}{7·9}$$
 $$= 3·8 \text{ m s}^{-2}$$

 $v = 25$ m s^{-1}
 $u = 0$ m s^{-1}
 $t = 7·2$ s
 $$a_{carB} = \frac{v - u}{t}$$
 $$= \frac{25 - 0}{7·2}$$
 $$= 3·47 \text{ m s}^{-2}$$

 ⇒ acceleration of car A is greater (than
 acceleration of car B)

 (*In this question it is not necessary to round the
 accelerations to the correct number of significant
 figures (2). The accelerations are not final answers.*)

 b) i) $a = ?$
 $v = 0$ m s^{-1}
 $u = 24$ m s^{-1}
 $t = 15$ s
 $$\text{braking } a_{carA} = \frac{v - u}{t}$$
 $$= \frac{0 - 24}{15}$$
 $$= -1·6 \text{ m s}^{-2}$$

 $v = 0$ m s^{-1}
 $u = 20$ m s^{-1}
 $t = 12$ s
 $$\text{braking } a_{carB} = \frac{v - u}{t}$$
 $$= \frac{0 - 20}{12}$$
 $$= -1·67 \text{ m s}^{-2}$$

 ⇒ Car B has more effective brakes (*larger
 negative acceleration*).

 ii) To make the test fair both cars should be
 brought to rest from the same initial speed
 (the only difference between the tests should
 be the braking systems of the two cars).

5.4 Force, gravity and friction

Page 89

1 a) Example answers: boy using force of muscles to
 stretch an elastic band or fold a piece of paper;
 weight of a person squashing a cushion when
 sitting down; wind force bending a tree, etc.

 b) Example answers: force of a car engine
 causing car to get faster; force of gravity
 making an object fall faster; girl using force
 of muscles to kick a stationary ball, etc.

 c) Example answers: man using force of (neck)
 muscles to head a ball; woman using muscle

force to hit hockey ball with a stick; gust of wind catching a falling leaf, etc.

2 $W = ?$ $\quad\quad\quad\quad\quad\quad W = mg$
   $m = 1220\,kg$ $\quad\quad\quad\quad\quad = 1220 \times 9{\cdot}8$
   $g = 9{\cdot}8\,N\,kg^{-1}$ $\quad\quad\quad = 11956$
   $\quad\quad\quad\quad\quad\quad\quad\quad = 12\,000\,N\ (= 12\,kN)$

3 $\quad m = 2130\,kg$
   $g_{Earth} = 9{\cdot}8\,N\,kg^{-1}$
   $g_{Mars} = 3{\cdot}7\,N\,kg^{-1}$
   $W_{Earth} = 2130 \times 9{\cdot}8 = 20\,874\,N$
   $W_{mars} = 2130 \times 3{\cdot}7 = 7881\,N$
   $\Delta W = 20\,874 - 7881 = 12\,993 = 1{\cdot}3 \times 10^4\,N$

4 a) For example: air resistance is not helpful when an aeroplane is taking off. The force of air resistance is reduced by making the shape of the aeroplane streamlined.
   b) For example: at the start of a race an athlete needs a good grip on the ground so that her feet do not slip. The frictional force is increased by making the sole of the running shoe ridged. (For sprinters, sometimes the running shoes have spikes!)

5.5 Newton's Laws

Page 95

1 a) Balanced – the motion of the girl is not changing.
   b) Unbalanced – the direction of the cyclist is changing.
   c) Unbalanced – the speed of the aeroplane is increasing.
   d) Unbalanced – the speed of the car is decreasing.
   e) Balanced – the electron is at rest.
   f) Unbalanced – the direction of the electron is changing.

2 a) Any stationary object; any object moving at a constant speed in a straight line.
   b) Any situation where the speed or direction of motion of an object (or both) is changing.

3 a) Forward forces: engine force. Backward forces: air resistance, force of friction between tyres and road.
   Downward forces: weight of car. Upward forces: forces between ground and tyres.

b) Forward forces: none. Backward forces: air resistance, force due to brakes, friction between tyres and road.
   Downward forces: weight of car. Upward forces: forces between ground and tyres.
   c) i) Horizontal forces are unbalanced. Vertical forces are balanced.
      ii) Horizontal forces are unbalanced. Vertical forces are balanced.

4 a) Resultant force, $F = 8300 - 3200 = 5100\,N$
   b) $m = 3{\cdot}4 \times 10^4\,kg$ $\quad\quad\quad F = ma$
      $F = 5100\,N$ $\quad\quad\quad \Rightarrow 5100 = 3{\cdot}4 \times 10^4 \times a$
      $a = ?$ $\quad\quad\quad\quad\quad\quad \Rightarrow \quad a = 0{\cdot}15\,m\,s^{-2}$

5 First find the resultant of the two applied forces.

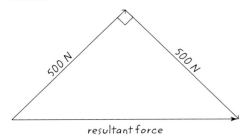
resultant force

Using Pythagoras' theorem, $F_{resultant}{}^2 = 500^2 + 500^2$
   $\Rightarrow F_{resultant} = 707{\cdot}1$
   $\quad\quad\quad\quad\quad = 707\,N$

The crate is moving at a constant speed so the forces acting on the crate are balanced.

$\Rightarrow F_{friction}$ between base of crate and factory floor $= (-)707\,N$

6 $m = 1200\,kg$ $\quad\quad\quad\quad\quad F = ma$
   $a = 0{\cdot}70\,m\,s^{-2}$ $\quad\quad\quad\quad = 1200 \times 0{\cdot}70$
   $F = ?$ $\quad\quad\quad$ Unbalanced force $= 840\,N$

7 $a = 500\,mm\,s^{-2} = 0{\cdot}5\,m\,s^{-2}$ (*Remember to check units!*)
   $F = 2{\cdot}0\,N$ $\quad\quad\quad\quad\quad\quad F = ma$
   $m = ?$ $\quad\quad\quad\quad\quad \Rightarrow 2{\cdot}0 = m \times 0{\cdot}5$
   $\quad\quad\quad\quad\quad\quad \Rightarrow$ mass of trolley $= 4{\cdot}0\,kg$

8 First find a:

$t = 5.0\,s$

$v = 6.4\,m\,s^{-1}$

$u = 2.4\,m\,s^{-1}$

Now use F = ma

Unbalanced force = 64 N

$a = ?$

$a = \dfrac{v - u}{t}$

$= \dfrac{6.4 - 2.4}{5.0}$

$\Rightarrow a = 0.80\,m\,s^{-2}$

$m = 80\,kg$

$\Rightarrow F = 80 \times 0.80$

$= 64\,N$

9 Let the mass of the normal car be m kg, so mass of less powerful car = $0.98m$ kg.

Let the force of the normal engine = F N, so force of less powerful car = $0.95F$ N.

Acceleration of normal car = $\dfrac{F}{m}$

Acceleration of less powerful car = $\dfrac{0.95F}{0.98m} = 0.969\dfrac{F}{m}$

$\Rightarrow$ less powerful car has 97% of the acceleration of the normal model.

10 Action-reaction forces act on different objects. Equal and opposite forces can cancel each other only if they act on the same object.

5.6 Energy, work and power

Page 99

1 $E_w = 450\,J$

$F = 30\,N$

$E_w = Fd$

$\Rightarrow 450 = 30 \times d$

$\Rightarrow \quad d = 15\,m$

2 Potential energy of metal sphere changes to kinetic energy as sphere swings. Maximum velocity of sphere occurs at lowest point of swing.

Assume kinetic energy gained = potential energy lost.

$\Rightarrow \dfrac{1}{2}mv^2 = mgh \Rightarrow v^2 = 2gh$ (*m cancels from both sides of the equation.*)

$h = 35\,mm = 0.035\,m$

$g = 9.8\,N\,kg^{-1}$

$\Rightarrow v^2 = 2 \times 9.8 \times 0.035$

$\Rightarrow v = 0.828 = 0.83\,m\,s^{-1}$

3 a) $P = 12\,000\,W$

$t = 15\,min = 900\,s$

$P = \dfrac{E}{t}$

$\Rightarrow E = 12\,000 \times 900$

$\Rightarrow E = 10\,800\,000$

$= 1.1 \times 10^7\,J$

b) $v = 6.0\,m\,s^{-1}$

$t = 15\,min = 900\,s$

$d = vt$

$\Rightarrow d = 6.0 \times 900$

$= 5400\,m$

c) Assume useful energy out of boat engine = work done by engine force.

$F = 850\,N$

$d = 5400\,m$

$E_w = Fd$

$= 850 \times 5400$

$= 4.59 \times 10^6$

$= 4.6 \times 10^6\,J$

d) Some of the energy used by the engine has been used to overcome forces of air resistance and water resistance.

e) The 'missing' energy has been converted to heat in the air, heat in the water and heat in the hull of the boat.

(*Do not include sound in your answer to this type of question. While some energy is converted to sound, the quantity is not big enough to make a difference to the calculated values of energy in and useful energy out.*)

4 Percentage efficiency is calculated from useful energy (or useful power) out divided by energy (or power) in. When energy is divided by energy (or power by power) the units cancel.

5 First calculate the potential energy of the ball before it is dropped.

$m = 0.7\,kg$

$g = 9.8\,m\,s^{-2}$

$h = 3.0\,m$

$E_p = mgh$

$= 0.7 \times 9.8 \times 3.0$

$= 20.58\,J$

Now assume that all of the potential energy is converted to kinetic energy.

$\Rightarrow E_k = \dfrac{1}{2}mv^2 = 20.58$

$0.5 \times 0.7 \times v^2 = 20.58$

$\Rightarrow v^2 = 58.8$

$\Rightarrow v = 7.668$

$= 7.7\,m\,s^{-1}$

$\Rightarrow$ Ball hits the ground at $= 7.7\,m\,s^{-1}$

Alternative solution: start with the assumption that $\dfrac{1}{2}mv^2 = mgh$

Mass cancels from both sides

$\Rightarrow \dfrac{1}{2}v^2 = gh$

$\Rightarrow v^2 = 2gh$

$\Rightarrow v^2 = 2 \times 9.8 \times 3.0$

$= 58.8$

$v = 7.7\,m\,s^{-1}$

(*For both methods, remember to take the square root.*)

6 The only difference between this and the previous question is the mass of the ball.

Start with the assumption that $\frac{1}{2}mv^2 = mgh$

Mass cancels from both sides

$\Rightarrow \frac{1}{2}v^2 = gh$
$\Rightarrow v^2 = 2gh$
$\Rightarrow v^2 = 2 \times 9\cdot8 \times 3\cdot0$
$\quad\quad = 58\cdot8$
$\Rightarrow v = 7\cdot7\,\text{m s}^{-1}$

7 First, calculate the time from throw till the rock reaches its highest point.

$u = 6\cdot0\,\text{m s}^{-1}$ $\quad\quad a = \dfrac{v - u}{t}$

$v = 0\cdot0\,\text{m s}^{-1}$ $\quad\quad \Rightarrow 9\cdot8 = \dfrac{6\cdot0}{t}$

$g = 9\cdot8\,\text{m s}^{-2}$ $\quad\quad \Rightarrow t = 0\cdot612\,\text{s}$

Now calculate the average velocity of the rock on its upward journey.

As the acceleration is uniform we can use

$\bar{v} = \dfrac{u + v}{2}$

$\Rightarrow \bar{v} = \dfrac{6\cdot0 - 0\cdot0}{2} = 3\cdot0\,\text{m s}^{-1}$

Now calculate vertical distance risen using
$d = \bar{v} \times t$
$\quad = 3\cdot0 \times 0\cdot612$
$\quad = 1\cdot836$
$\Rightarrow$ maximum height $= 1\cdot8\,\text{m}$

8 Assume $E_k = E_p$ and start from $v^2 = 2gh$

$g = 9\cdot8\,\text{m s}^{-2}$ $\quad\quad \Rightarrow v^2 = 2 \times 9\cdot8 \times 0\cdot45$
$h = 0\cdot45\,\text{m}$ $\quad\quad\quad = 8\cdot82$
$\quad\quad\quad\quad\quad\quad\quad \Rightarrow v = 2\cdot969$
$\Rightarrow$ velocity at lowest point $= 3\cdot0\,\text{m s}^{-1}$

9 a) $h = 1\cdot0\,\text{m}, v^2 = 2gh \Rightarrow v^2 = 2 \times 9\cdot8 \times 1\cdot0 = 19\cdot6 \Rightarrow v = 4\cdot427 = 4\cdot4\,\text{m s}^{-1}$

b) $h = 2\cdot0\,\text{m}, v^2 = 2gh \Rightarrow v^2 = 2 \times 9\cdot8 \times 2\cdot0 = 39\cdot2 \Rightarrow v = 6\cdot26 = 6\cdot3\,\text{m s}^{-1}$

c) $h = 3\cdot0\,\text{m}, v^2 = 2gh \Rightarrow v^2 = 2 \times 9\cdot8 \times 3\cdot0 = 58\cdot8 \Rightarrow v = 7\cdot668 = 7\cdot7\,\text{m s}^{-1}$

d) $h = 4\cdot0\,\text{m}, v^2 = 2gh \Rightarrow v^2 = 2 \times 9\cdot8 \times 4\cdot0 = 78\cdot4 \Rightarrow v = 8\cdot854 = 8\cdot9\,\text{m s}^{-1}$

e) $h = 5\cdot0\,\text{m}, v^2 = 2gh \Rightarrow v^2 = 2 \times 9\cdot8 \times 5\cdot0 = 98\cdot0 \Rightarrow v = 9\cdot89 = 10\,\text{m s}^{-1}$

10

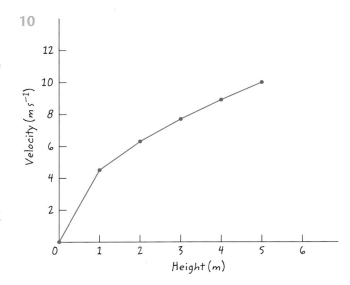

11 a) $m = 0\cdot45\,\text{kg}$ $\quad\quad E_p = mgh$
$h = 1\cdot2\,\text{m}$ $\quad\quad\quad\quad = 0\cdot45 \times 9\cdot8 \times 1\cdot2$
$g = 9\cdot8\,\text{m s}^{-2}$ $\quad\quad\quad = 5\cdot292$
$\quad\quad\quad\quad\quad\quad\quad\quad = 5\cdot3\,\text{J}$

b) $v = 3\cdot5\,\text{m s}^{-1}$ $\quad\quad E_k = \frac{1}{2}mv^2$
$\quad\quad\quad\quad\quad\quad\quad = 0\cdot5 \times 0\cdot45 \times 3\cdot5^2$
$\quad\quad\quad\quad\quad\quad\quad = 2\cdot756\,\text{J}$

Energy 'lost' due to friction $= 2\cdot5\,\text{J}$

c) $v^2 = 2gh$
$\quad = 2 \times 9\cdot8 \times 1\cdot2$
$\quad = 23\cdot52$
$\Rightarrow v = 4\cdot849 = 4\cdot8\,\text{m s}^{-1}$

12 a) $m = 2\cdot0\,\text{kg}$ $\quad\quad E_p = mgh$
$h = 1\cdot5\,\text{m}$ $\quad\quad\quad\quad = 2\cdot0 \times 9\cdot8 \times 1\cdot5$
$g = 9\cdot8\,\text{m s}^{-2}$ $\quad\quad\quad = 29\cdot4 = 29\,\text{J}$

b) Clock uses $0\cdot083\,\text{J}$ per hour – number of hours operation $= \dfrac{29\cdot4}{0\cdot083} = 354\cdot21$

Clock will operate for $= 350$ hours

c) The potential energy is converted to **kinetic energy** of the parts of the clock mechanism.

13 $m = 1\cdot1\,\text{kg}$ Loss of potential energy $= mg(h_1 - h_2)$
$h_1 = 1\cdot5\,\text{m}$ $\quad\quad\quad\quad = 1\cdot1 \times 9\cdot8 \times (1\cdot5 - 1\cdot3)$
$h_2 = 1\cdot3\,\text{m}$ $\quad\quad\quad\quad = 2\cdot156$
$\quad\quad\quad\quad\quad\quad\quad\quad = 2\cdot2\,\text{J}$

14 The pendulum uses energy to cut through the air/push the air out of its path and this slowly reduces the speed of the pendulum.

15 There is no more driving force and the kinetic energy is 'used' to push the air out of its path and turn the wheels, tyres and other components.

5.7 Projectiles
Page 103

1 First calculate the speed of the stone when it reaches the water surface:

$u = 0\,\text{m s}^{-1}$
$g = 9.8\,\text{m s}^{-2}$
$t = 3.6\,\text{s}$

$a = \dfrac{v - u}{t}$

$\Rightarrow 9.8 = \dfrac{v - 0}{3.6}$

$\Rightarrow v = 35.3 = 35\,\text{m s}^{-1}$

Now sketch the speed-time graph for the stone – a rough sketch is okay.

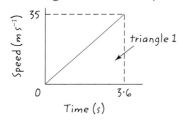

Now calculate the distance fallen by the stone:

Distance = area of triangle 1

$= (\dfrac{1}{2} \times 3.6 \times 35.3) = 63.54\,\text{m}$

$\Rightarrow$ water surface is 64 m below the top of the well.

2 First calculate the time taken for the rocket to reach its highest point:

$u = -60\,\text{m s}^{-1}$
$a = 9.8\,\text{m s}^{-2}$
$v = 0\,\text{m s}^{-1}$

$a = \dfrac{v - u}{t}$

$\Rightarrow 9.8 = \dfrac{0 - (-60)}{t}$

$\Rightarrow t = 6.12 = 6.1\,\text{s}$

Ignoring air resistance, the 'up' and 'down' parts of the rocket's motion take the same time, so time for rocket to fall from its highest point to Earth = 6.1 s.

Now sketch the velocity–time graph for the falling rocket – again, a rough sketch is okay.

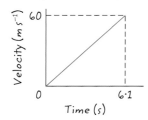

Note: the final velocity is $+60\,\text{m s}^{-1}$, i.e. the direction is downwards.

Distance fallen by rocket = area under velocity–time graph

$= \dfrac{1}{2} \times 6.12 \times 60 = 183.6\,\text{m}$

$\Rightarrow$ Rocket reaches a height of 184 m

3 Ignoring air resistance, the horizontal velocity of the object is constant. In each second the horizontal distance travelled by the object is the same.

The initial vertical velocity of the object is zero and the vertical acceleration is $9.8\,\text{m s}^{-2}$. This means that each second the vertical velocity increases by $9.8\,\text{m s}^{-1}$. After 1 s the vertical velocity is $9.8\,\text{m s}^{-1}$, after 2 s it is $19.6\,\text{m s}^{-1}$ and so on. In each consecutive second the vertical distance travelled by the object increases.

4 a)

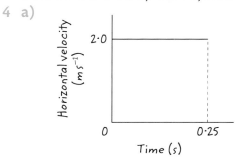

b) Now consider vertical motion.

$t = 0.25\,\text{s}$
$a = 9.8\,\text{m s}^{-2}$
$u = 0\,\text{m s}^{-1}$

$a = \dfrac{v - u}{t}$

$\Rightarrow 9.8 = \dfrac{v - 0}{0.25}$

$\Rightarrow v = 2.45\,\text{m s}^{-1}$

c) Horizontal velocity is contant $\Rightarrow d = vt$
$v = 2.0\,\text{m s}^{-1}$
$t = 0.25\,\text{s}$
$\Rightarrow d = 2.0 \times 0.25$
$= 0.50\,\text{m}$

d) First calculate average vertical velocity.
$v = 2.45\,\text{m s}^{-1}$

$\overline{v} = \dfrac{u + v}{2}$ (See the third 'Hints and tips' box on page 76.)

$u = 0\,\text{m s}^{-1}$

$t = 0.25\,\text{s}$

$\Rightarrow \bar{v} = \dfrac{0 + 2.45}{2}$

$= 1.225\,\text{m s}^{-1}$

$d = \bar{v}t$

$= 1.225 \times 0.25$

$= 0.306 = 0.31\,\text{m}$

5 a) First use horizontal motion where v is constant at $4.8\,\text{m s}^{-1}$.

$v = 4.8\,\text{m s}^{-1}$

$d = 12\,\text{m}$

$d = vt$

$\Rightarrow 12 = 4.8 \times t$

$\Rightarrow t = 2.5\,\text{s}$

   b) Now consider vertical motion.

$t = 2.5\,\text{s}$

$u = 0\,\text{m s}^{-1}$

$a = \dfrac{v - u}{t}$

$\Rightarrow 9.8 = \dfrac{v - 0}{2.5}$

$\Rightarrow v = 24.5 = 25\,\text{m s}^{-1}$

6 It is not possible to calculate the horizontal distance travelled by the ball using only time, an initial vertical velocity of $0\,\text{m s}^{-1}$ and the value of g. These pieces of information can be used to work out the initial height of the tennis ball. The horizontal motion of the ball has no effect on the vertical motion – the time for the ball to reach the ground is the same for any value of horizontal velocity.

7 The vertical forces acting on the parachutist are her weight and upward air resistance. Her weight is constant throughout her fall.

Initially, vertical $F_{\text{air resistance}} = 0\,\text{N} \Rightarrow F_{\text{unbalanced}} = $ weight $\Rightarrow a = 9.8\,\text{m s}^{-2}$.

As her speed increases, $F_{\text{air resistance}}$ increases $\Rightarrow F_{\text{unbalanced}}$ decreases $\Rightarrow a$ decreases.

When $F_{\text{air resistance}} = $ weight she falls at a constant speed $\Rightarrow$ final $a = 0\,\text{m s}^{-2}$.

8 The feather and hammer hit the surface of the moon at the same time. The moon does not have an atmosphere so there is no air resistance to reduce the acceleration of the feather.

6

Space

6.1 Space exploration

Page 111

1 a) $E_p = mgh$

$= 18\,000 \times 9.8 \times 100\,000$

$= 1.76 \times 10^{10}$

$= 1.8 \times 10^{10}\,\text{J}$

$E_k = \dfrac{1}{2}mv^2$

$E_k = 0.5 \times 18\,000 \times 5000 \times 5000$

$= 2.25 \times 10^{11}\,\text{J}$

   b) Heat energy

   c) It must be able to withstand high temperature and pressures and be able to radiate some of the heat away.

2 Any three, such as: global position systems, mobile communications, weather forecasting, scientific research and study, television channels, etc.

3 Student answers will vary.
4 Student answers will vary.

6.2 Cosmology

Page 115

1 Stars and galaxies emit many types of radiation and using different types of detectors gives us more information about that star.

2 Line spectra give information relating to the chemical composition of that star.

3 $4.2 \times 300\,000\,000 \times 365.25 \times 24 \times 60 \times 60 = 3.98 \times 10^{16}\,\text{m}$

4 It would take 8.6 years for the data to reach us and it would take 8.6 years for any change to its programming to reach the probe.

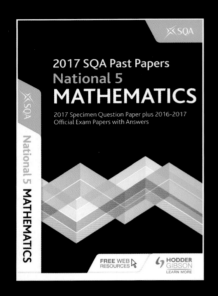

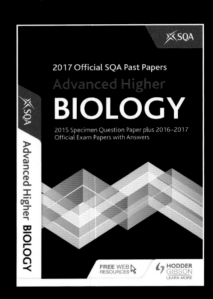

How to Pass

NATIONAL 5
Art & Design

Includes extensive Portfolio advice

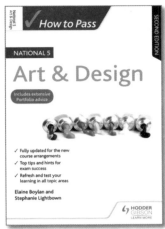

✓ Fully updated for the new course arrangements
✓ Top tips and hints for exam success
✓ Refresh and test your learning in all topic areas

Elaine Boylan and Stephanie Lightbown

HODDER GIBSON
LEARN MORE

How to Pass

NATIONAL 5
Biology

✓ Fully updated for the new course arrangements
✓ Top tips and hints for exam success
✓ Refresh and test your learning in all topic areas

Billy Dickson and Graham Moffat

HODDER GIBSON
LEARN MORE

How to Pass

NATIONAL 5
Business Management

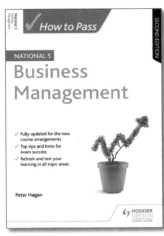

✓ Fully updated for the new course arrangements
✓ Top tips and hints for exam success
✓ Refresh and test your learning in all topic areas

Peter Hagan

HODDER GIBSON
LEARN MORE

How to Pass

NATIONAL 5
Chemistry

✓ Fully updated for the new course arrangements
✓ Top tips and hints for exam success
✓ Refresh and test your learning in all topic areas

Barry McBride

HODDER GIBSON
LEARN MORE

How to Pass

NATIONAL 5
Computing Science

✓ Fully updated for the new course arrangements
✓ Top tips and hints for exam success
✓ Refresh and test your learning in all topic areas

David Alford

HODDER GIBSON
LEARN MORE

How to Pass

NATIONAL 5
English

✓ Fully updated for the new course arrangements
✓ Top tips and hints for exam success
✓ Refresh and test your learning in all topic areas

David Swinney

HODDER GIBSON
LEARN MORE

How to Pass

NATIONAL 5
French

✓ Fully updated for the new course arrangements
✓ Top tips and hints for exam success
✓ Refresh and test your learning in all topic areas

Douglas Angus

HODDER GIBSON
LEARN MORE

How to Pass

NATIONAL 5
Geography

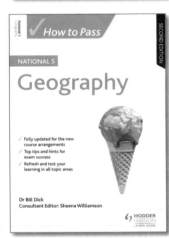

✓ Fully updated for the new course arrangements
✓ Top tips and hints for exam success
✓ Refresh and test your learning in all topic areas

Dr Bill Dick
Consultant Editor: Sheena Williamson

HODDER GIBSON
LEARN MORE

How to Pass

NATIONAL 5
German

✓ Fully updated for the new course arrangements
✓ Top tips and hints for exam success
✓ Refresh and test your learning in all topic areas

Kirsten Herbst-Gray

HODDER GIBSON
LEARN MORE

How to Pass

NATIONAL 5
History

✓ Fully updated for the new course arrangements
✓ Top tips and hints for exam success
✓ Refresh and test your learning in all topic areas

John A. Kerr and Jerry Teale

HODDER GIBSON
LEARN MORE

How to Pass

NATIONAL 5
Applications of Mathematics

✓ Fully updated for the new course arrangements
✓ Top tips and hints for exam success
✓ Refresh and test your learning in all topic areas

Mike Smith

HODDER GIBSON
LEARN MORE

How to Pass

NATIONAL 5
Maths

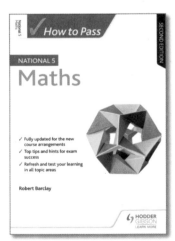

✓ Fully updated for the new course arrangements
✓ Top tips and hints for exam success
✓ Refresh and test your learning in all topic areas

Robert Barclay

HODDER GIBSON
LEARN MORE

How to Pass

NATIONAL 5
Modern Studies

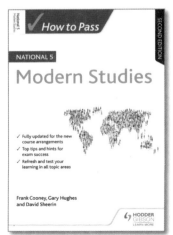

✓ Fully updated for the new course arrangements
✓ Top tips and hints for exam success
✓ Refresh and test your learning in all topic areas

Frank Cooney, Gary Hughes and David Sheerin

HODDER GIBSON
LEARN MORE

How to Pass

NATIONAL 5
Music

✓ Fully updated for the new course arrangements
✓ Top tips and hints for exam success
✓ Refresh and test your learning in all topic areas

Joe McGowan

HODDER GIBSON
LEARN MORE

How to Pass

NATIONAL 5
Physics

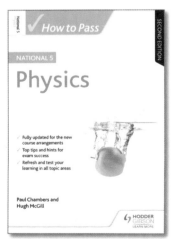

✓ Fully updated for the new course arrangements
✓ Top tips and hints for exam success
✓ Refresh and test your learning in all topic areas

Paul Chambers and Hugh McGill

HODDER GIBSON
LEARN MORE

SECOND EDITION

Physics

✓ How to Pass

NATIONAL 5

Physics

Get your best grade with this updated edition of How to Pass National 5 Physics.

This book contains all the advice and support you need to revise successfully for your National 5 exam. It has been updated to reflect the most recent changes to the course specifications, ensuring you know exactly what to expect in the final exam. It combines an overview of the course syllabus with advice from top experts on how to improve exam performance, so you have the best chance of success.

There are *How to Pass Guides* and *Practice Exam Papers* available in many more subjects at National 5 and Higher. Check them out at our web address below!

Track your knowledge with regular checks on what you should know

Illustrated throughout to aid comprehension

Key points are highlighted throughout

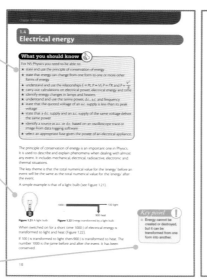

Helpful tips reinforce classroom learning a underline critical poi

Examples help to contextualise learning

Exam-style questions provide essential practice

Hodder Gibson
Scotland's Number One Educational Publisher

At Hodder Gibson we offer a wide variety of textbooks, revision guides and digital educational resources for the Scottish curriculum.

HODDER GIBSON
t: 01235 827827
e: education@bookpoint.co.uk
w: hoddergibson.co.uk

£11.

ISBN 978-1-510-42105-9

9 781510 421059

FSC